Rapid Modeling Solutions:

Introduction to Simulation and Simio

C. Dennis Pegden

David T. Sturrock

504 Beaver St, Sewickley, PA 15143, USA

www.simio.com

Rapid Modeling Solutions:

Introduction to Simulation and Simio

By C. Dennis Pegden and David T. Sturrock of Simio LLC

First Edition – Last updated October 10, 2013 – Economy (Black & White Interior)

Copyright ©2014 Simio LLC. All rights reserved.

This book was written to provide an introduction to simulation and Simio. We encourage its use as a companion to the Simio training/evaluation software which can be downloaded at no charge from www.simio.com/products.

We welcome feedback and contributions to textbook@simio.com.

Contents

Table of Contents

Chapter 1: Introduction to Simulation 1
    Overview 1
    What is Simulation? 1

Chapter 2: Getting Started with Simio 2
    Overview 2
    Models and Projects 3
    The Simio User Interface 3
    Objects and Libraries 5

Chapter 3: Introduction to the Standard Library 6
    Overview 6
    Source and Sink 9
    Server 10
    Workstation 11
    Resource 14
    Worker 15
    Combiner and Separator 15
    Vehicle 16
    Basic and Transfer Nodes 18
    Connector, Path and TimePath 20
    Conveyor 21
    ModelEntity* 22

Chapter 4: Our First Model 23
    Overview 23
    Manipulating Facility Views 23
    Editing Object Properties 24
    A Simple Flow Line Model 25
    Defining Model Properties and Experiments 28
    Interpreting the Results 30
    Summary 32

Chapter 5: Network Travel 33
    Overview 33
    Entities and Attributes 33
    Networks 35
    Node Routing Logic 36

Example: Routing by Link Weights 36
Setting the Entity Destination 38
Example: Select From List for Dynamic Routing 39
Example: By Sequence 41
Summary 43

Chapter 6: More about the Standard Library 44
Overview 44
Preliminary Concepts 44
Source and Sink 45
Server 48
Combiner and Separator 51
Example: Combine then Separate 52
Basic Node and Transfer Node 53
Connector, TimePath, and Path 54
Example: Bidirectional Paths 54
Conveyors 55
Example: Merging Conveyors 56
Vehicle 57
Example: On-Demand Pickups 58
Workstation 59
Worker 60
Example: Moveable Operators 61
Summary 62

Chapter 7: Data Driven Models 63
Overview 63
Data Tables 63
Example: Multiple Entities from a Single Source 65
Example: Product Routings using Separate Sequence Tables 67
Example: Product Routings using a Table Relation 69
Summary 70

Chapter 8: Processes 71
Overview 71
Processes 71
Process Types 72
Building Processes 74
Steps and Associated Elements 75
Example: Using a Moveable Resource with Server Failures (Seize/Release) 79
Example: Initializing an Array using a Table 81
Summary 84

Chapter 9: Object Definitions 85

- Overview 85
- Basic Concepts 85
- External View 85
- Model Logic 87
- Properties, States, and Events 88
- Example: Tandem Server 89
- Example: A Base Lathe 92
- Example: Server with Repairman 94
- Summary 97

Glossary 98

More Information 103
- Technical Support 103
- More Information 103
- Using This Material in Support of Teaching 103

Appendix 1: Simio and Simulation: Modeling, Analysis, Applications 104

Chapter 1: Introduction to Simulation

Overview

Welcome to the world of simulation. Simulation provides a unique way to examine the future and make intelligent decisions based on what you learn. While simulation technology has been around for decades, it is still rapidly evolving. Advances in object-oriented approaches provide rapid modeling and the flexibility to model complex systems that could not be modeled just a few years ago. And integrated 3D animation makes the creation of compelling 3D visualizations easy, and this in turn helps assure more robust, understandable models and better communication with stakeholders.

The team of architects behind Simio have been leaders in simulation since the early 1980's, playing key roles in the design and development of four previous market-leading products. Simio is the result of this team applying their collective 180 years of simulation experience and using the very latest in technology and development techniques to create a new generation of simulation problem solving capability.

What is Simulation?

While the tools are becoming steadily easier to use, there is much more to successful simulation then just using the best available tool. It starts with knowing what simulation can do for you and how to effectively use that power. By special arrangement, we have been allowed to include Chapter 1 of *Simio and Simulation: Modeling, Analysis, Applications* (SASMAA) in an appendix to this book.

SASMAA starts with an overview of what simulation is, the different types of simulation, and a survey of some common applications. Then it describes the role of randomness and stochastics in simulation, the simulation process, and introduces the concepts of input analysis, output analysis, and model verification and validation. Then it discusses when you should and should not consider using simulation for a project.

SASMAA Chapter 1 finishes with a valuable discussion of Simulation Success Skills, identifying opportunities to exploit, and pitfalls to avoid, to help you make every project a success. This includes such topics as setting project objectives, proper use of a functional specification, the critical importance of an iterative project approach, and project management with the right level of agility. This chapter ends by presenting a *Stakeholder Bill of Rights* and a *Simulationist Bill of Rights* – rules to live by to help assure project success.

As of this writing SASMAA is currently in its second edition. The appendix contains an early look at the upcoming third edition. You are strongly encouraged to obtain the entire book, available in both printed and e-book form, for a much more thorough coverage of Simio and simulation.

If you really want to just start building models, you can proceed directly to Chapter 2. But even if you are already familiar with simulation, and especially if you are not, **you are strongly encouraged to study Appendix 1 – *Introduction to Simulation* first.** Go ahead and read it now … we'll wait here until you get back.

Chapter 2: Getting Started with Simio

Overview

The Simio modeling software lets you build and run dynamic 3D animated models of a wide range of systems – e.g. factories, supply chains, emergency departments, airports, and service systems. Simio is a family of products comprised of the ***Training/Evaluation, Express, Design, Team***, and ***Enterprise*** Editions.

- The Training Evaluation Edition is a free download that allows unlimited model-building, but some features like Save and Experimentation are limited to use with only small models.

- The Express Edition allows a modeler to get up to speed quickly building object-oriented models using Simio's Standard Object Library. Fully functional 3D modeling and animation capabilities are included as described in Chapters 1 through 4 of this book.

- The Design Edition is our standard product offering. This version incorporates all the functionality of the Express Edition, plus provides the capability to customize object behavior with add-on process-oriented logic as described in Chapter 8, and provides the ability to create and distribute your own custom libraries as described in Chapter 9.

- The Team Edition is a special version of the Design Edition that lets you build and distribute models using the freely available Simio Evaluation version as a runtime platform. Models built with the Team Edition and above will run and generate results with the Training Evaluation Edition[1]. This version is ideal for consultants that want to deliver a running model to their customer without requiring their customer to purchase Simio. The Team Edition also provides support for distributing run executions across a workgroup of computers.

- And finally, the Enterprise version is our top of the line product that includes extra general modeling features as well as scheduling specific features that allow it to be used for Risk-based Planning and Scheduling (RPS) applications.

Simio employs an object approach to modeling, whereby models are built by combining objects that represent the physical components of the systems. An object has its own custom behavior as defined by its internal model that responds to events in the system. For example a production line model is built by placing objects that represent machines, conveyors, forklift trucks, aisles, etc. You can build your models using the objects provided in the Standard Object Library, which is general purpose set of objects that comes standard with Simio. You can also build your own libraries of objects that are customized for specific application areas. You can also modify and extend the Standard Library object behavior using process logic.

An **object** (or **model**) is defined by its properties, states, events, external view, and logic. These are key Simio concepts to understand for building and using objects.

Properties are input values that can be specified by the user of the object. For example an object representing a server might have a property that specifies the service time. When the user places the server object into their facility model they would also specify this property value.

---

[1] Some limitations apply.

An object's **states** are dynamic values that may change as the model executes. For example the busy and idle status of a server object could be maintained by a state variable named Status that is changed by the object each time it starts or ends service on a customer.

Events are things that the object may "fire" at selected times. For example a server object might have an event fire each time the server completes processing of a customer, or a tank object might fire an event whenever it reaches full or empty. Events are useful for informing other objects that something important has happened.

The external view of an object is the 3D graphical representation of the object. This is what a user of the object will see when it is placed in their facility model.

The object's logic is an internal model that defines how the object responds to specific events that may occur. For example a server object may have a model that specifies what actions take place when a customer arrives to the server. The internal model gives the object its unique behavior.

Models and Projects

When you first open Simio, you see a new model open within a project. Notice that the Support ribbon includes links to the Simio Reference Guide, Training Videos, Example models, and SimBits. The SimBits are small searchable models that illustrate how to approach common modeling situations. Under Training Videos you will find three complete training series including one which roughly parallels this book. All are available at no charge.

Models are defined within a project. A project may contain any number of models and associated experiments (discussed later). A project will typically contain a main model and an entity model. When you open up a new project, Simio automatically adds the main model (default name Model) and entity model (default name ModelEntity) to the project. You can rename the project and these models by right clicking on them in the project navigation tree. You can also add additional models to the project by right clicking on the project name. This is typically done to create sub-models that are then used in building the main model.

The entity model is used to define the behavior of the entities that move through the system. In older simulation systems entities cannot have behavior, and therefore there is no mechanism for building an entity model. However in Simio entities can have behaviors that are defined by their own internal model. The default model entity is "dumb" in that it has no explicit behavior, however as you will see later you can modify the entity model to take specific actions in response to events. You can also have multiple types of entity models in your project, with each having its behavior. For example in a model of an emergency department you could have different entities representing patients, nurses, and doctors.

As we will see later a project can also be loaded into Simio as modeling library. Hence some projects contain a collection of models for a specific application, and other projects contain models that are primarily used as building blocks for other models. The same project can either be opened for editing or be loaded as library.

The Simio User Interface

The initial view of your Simio project is shown below. The key areas in this screen include the ribbons across the top (currently showing the Run ribbon), the tabbed panel views with the Facility highlighted just below the ribbons, the libraries on the left, the browse panel on the right, and the Facility window in the center.

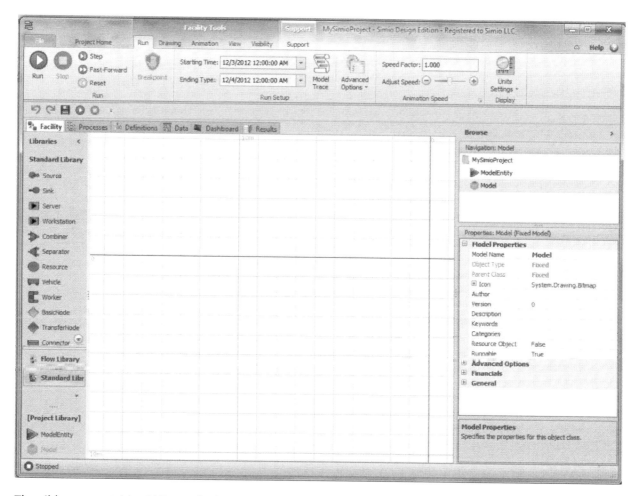

The ribbons are tabbed UI panels that provide a convenient way quickly to access the many functions available to you for building, animating, and running models. You can manually switch between ribbons by clicking on the ribbon tabs. You can also condense the ribbons to just the tabs by double-clicking on any tab. In this mode clicking on the tab will expand the ribbon until you click in the model, and then the ribbon will then shrink back to just tabs. Double-clicking again on any ribbon tab will restore the ribbons to their regular size.

The Browse panel on the right provides for project navigation and property editing. The upper navigation window is used to switch between the Project view, and the associated models and experiments. You can switch between models by simply clicking on the model in the navigation tree. For example clicking on the main model will make this the active model and display the Facility window for that model. The properties window that is located in the Browse panel immediately below the navigation window is used for editing the properties of objects. We will discuss this window in more detail later. You can open and close the Browse panel by clicking on the < and > in the title bar of the panel.

Whenever the Facility window of a model is selected the Libraries panel on the left displays the libraries that are open and available for modeling within the facility. The libraries will include the Standard Library, the Flow Library, the Project Library, and any additional projects that have been loaded as libraries from the Project Home ribbon. The Standard Library is a general purpose library of objects that is provided with Simio for modeling a wide range of systems. The Flow Library is a library of objects for modeling flow processing systems. The Project Library is a library of objects corresponding to the current models in your project. This lets you use your project models as sub-models that can be placed multiple times within a model. Note that the active model is grayed out in the Project Library since a model cannot be placed inside itself.

The Facility window that is shown in the center is drawing space for building your object-based model. This panel is shown whenever the Facility tab is selected for a model. This space is used to create both the object-based logic and the animation for the model in a single step. The other panel views associated with a model include Processes (for Design and Team Editions only), Definitions, Data, Dashboard, and Results. The Processes panel is used for defining custom process logic for your models. The ability to mix object-based and process modeling within the same model is one of the unique and very powerful features of Simio Design and Team Editions – this combines the rapid modeling capabilities of objects with the modeling flexibility of processes. The Definitions panel is used to define different aspects of the model, such as its external view and the properties, states, and events that are associated with the model. The Data panel is used to define data that may be used by the model, and imported/exported to external data sources. The Dashboard panel provides a 2D drawing space for placing buttons, dials, plots, etc., for real time viewing and interactions with the model. The Results panel displays the output from the model in the form of both a pivot grid as well as traditional reports. Note that you can view multiple model windows at the same time by dragging a window tab and dropping it on one of the layout targets. For now we will be focusing on the Facility window and will defer further discussion of these additional model windows until later.

Objects and Libraries

The objects within a library are one of five basic types:

- **Fixed**: Has a single fixed location in the system such as a machine.
- **Link**: Provides a pathway over which entities may move.
- **Node**: Defines an intersection between one or more incoming/outgoing links. Nodes can also be associated with fixed objects to provide entry and exit points for the object.
- **Entity**: Defines a dynamic object that can be created and destroyed, move over a network of links and nodes, and enter/exit fixed objects through their associated nodes.
- **Transporter**: Defines a special type of entity that can also pickup and drop off other entities at nodes.

Note that these types define the general behavior, but not the specific behavior of the object. The specific behavior an object is defined by the internal model for that object. For example we could have a library of a half a dozen different types of transporters, each with their own specific behaviors as defined by their models. However they would all share the common ability to move across a network of links and nodes, and pickup and drop off entities at nodes along the way.

A library may contain objects of any these five types. The Standard Library includes objects of all types except for the entity type, since the entity is typically defined within the Project Library.

Chapter 3: Introduction to the Standard Library

Overview

The Simio Standard Library contains 15 objects that will be the basis of most models. In this chapter we will provide a brief introduction to those objects, then in later chapter we will revisit them to explore even more of their functionality. The Standard Library contains the following objects:

Object	Description
Source	Generates entity objects of a specified type and arrival pattern.
Sink	Destroys entities that have completed processing in the model.
Server	Represents a capacitated process such as a machine or service operation.
Workstation	Models a complex workstation with setup, processing, and teardown phases and secondary resource and material requirements.
Resource	A generic object that can be seized and released by other objects.
Worker	A moveable resource that may be seized and released for tasks as well as used to transport entities between node locations.
Combiner	Combines multiple member entities together with a parent entity (e.g. a pallet).
Separator	Splits a batched group of entities or makes copies of a single entity.
Vehicle	A transporter that can follow a fixed route or perform on demand transport pickups/drop offs. Additionally, an 'On Demand' routing type vehicle may be used as a moveable resource that is seized and released for non-transport tasks.
BasicNode	Models a simple intersection between multiple links.
TransferNode	Models a complex intersection for changing destination and travel mode.
Connector	A simple zero-time travel link between two nodes.
Path	A link over which entities may independently move at their own speeds.
TimePath	A link that has a specified travel time for all entities.
Conveyor	A link that models both accumulating and non-accumulating conveyor devices.
ModelEntity*	An entity typically represents a part, person, or other object that is dynamically created, flows through the system, and then leaves the system. *Because entities are often customized they are not part of the Standard Library, but are instead automatically added to your project for easy revision.

Each object has properties that control its behavior; e.g. the Source has an inter-arrival time, the Path has a property to control entity passing, and the Server has a processing time. Let's take a brief look at these objects and their associated properties.

Simio provides a *Show Commonly Used Properties Only* checkbox at the top of the Properties Window.

☑ Show Commonly Used Properties Only

When this mode is enabled (the default), the display is limited to the key set of properties that defines the core behavior of each object. Many standard features such as failures, state assignments, secondary resource allocations, financials, and custom add-on process logic that are provided by the Standard Library objects are not displayed. This mode allows the beginner to focus on the key concepts for each object without getting mired down in additional complexities.

When this option is disabled (unchecked) the full set of properties for each object is exposed. Note that this mode does not directly affect model behavior or results. It is just hiding/un-hiding properties but not changing property values. This mode is typically used when first learning Simio and then disabled once the basic concepts are mastered. Initially will focus on the commonly used properties and resulting behavior, but we will explain the full set of properties in the chapters that follow.

We recommend leaving this mode checked at least while you are learning Simio. After you are more comfortable with the basic options or when you need to access more advanced features, you can uncheck this option to all display the full set of properties for each object.

Before describing the commonly used properties for each of the Standard Library objects, let's first briefly discuss the process of building a model in Simio using this library. To build a model you drag objects from the library and place them in your Facility View. The following shows a very simple example where we have placed a Source, Server, and Sink connected by Paths.

Introduction to Simio

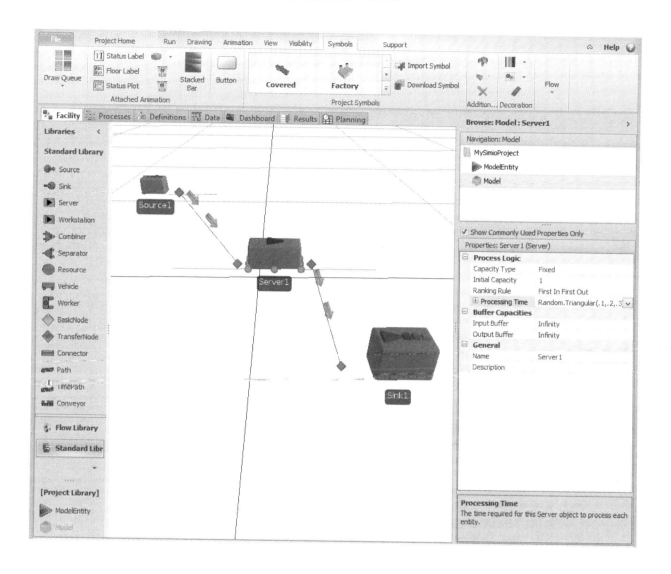

This might represent a simple service system such as a security check at an airport. Entities are created at the Source, move across the first Path to the Server, wait in line to be processed at the Server, and then move across the second Path to the Sink. In the example above we have clicked on the Server (highlighted by the green selection handles) which displays its basic properties in the Property Grid on the right. The properties are organized into categories that can be expanded and collapsed by clicking on the +/- to the left of the category name.

Learning to model in Simio begins with learning the objects in the Standard Library along with their associated properties that control their behavior, and then mastering the skill of combining these objects together to represent complex systems. Although a number of these objects have many different properties that control their behavior, the core behavior of each object is controlled by a small number of properties. The Commonly Used Properties mode allows us to focus our attention on these core properties.

Note that in our simple example the Source, Server, and Sink are connected together with Paths. A Path is a type of link that is used to define a network over which entities move. Links are drawn as a polyline connected to nodes that define locations within the model where links can start or end. Links and nodes combine together to form a network over which entities and transporters move. Nodes can either be free-standing or associated with an object. In this example the Source has an associated Output node, the Server has associated Input and Output nodes, and the Sink has an associated Input node. Associated nodes serve as potential entry and exit points for entities that enter and depart objects. When you place an object Simio also automatically places any associated nodes that have been defined for the object.

We will now describe the commonly used properties for each of the objects in the Standard Library.

Source and Sink

The **Source** is used to generate a sequence of arrivals to the system that exit the object at the Output node associated with the Source. These arrivals represent the dynamic entities that enter and leave the system and might correspond to customers, patients, jobs, phone calls, messages, etc. The commonly used properties on the Source are shown below:

- Entity Arrival Logic

Entity Type	The type of entities created by this Source object.
Arrival Mode	The mode used for generating a stream of entity arrivals.
Time Offset	The time of the first arrival.
Interarrival Time	The time interval between two successive arrival events.
Entities Per Arrival	The number of entities created for each arrival event.

- Stopping Conditions

Maximum Arrivals	The maximum number of arrival events to be generated.

- General

Name	The name of this object.
Description	A description of the use of this object in the model.

The first property for the source is the *Entity Type* and specifies the type of entity to create at this source. This property defaults to the generic *Default Entity* which is automatically created by Simio. You can also place additional entities into your model and then specify the *Entity Type* as one of these additional entities. The second property is the Arrival Mode and specifies the mode that is used to generate the sequence of entity arrivals. Each option for Arrival Mode switches in associated properties for that mode. The default mode is Interarrival Time with a *Time Offset* for the first arrival (defaulting to 0.0 to indicate immediate first arrival) and the arrival sequence based on an *Interarrival Time* property that is typically specified as a random distribution. This property defaults to random samples from an exponential distribution with a mean of .25 minutes. This distribution is commonly used to model random Poisson arrival processes such as customers arriving to a store. Other modes that are available include *Time Varying Arrival Rate* (random arrival pattern where the rate varies over time based on a Rate Table), *On Event* (creates an arrival each time a specified event occurs), and *Arrival Table* (generates a list of arrivals defined in a table). The *Entities per Arrival* property is used to specify the number of entities that arrive at each arrival event. This property defaults to 1, but can be changed to any valid expression. For example groups of varying size arriving to a restaurant could be modeled by specifying this property as a random variable. The Source also has a property named *Maximum Arrivals* that limits the number of arrival events at this source. This defaults to infinite but can be specified as any valid expression. The final properties for the Source (and all Standard Library objects) are *Name* and *Description*. Given each object both a meaningful name as well as describing what it does can be an important part of documenting your model and making it easier to understand.

The **Sink** is used to model the departure of entities from the system. When an entity enters the Sink at the associated Input node it is destroyed and removed from the model. The only commonly used properties on the Sink are *Name* and *Description*.

-General	
Name	The name of this object.
Description	A description of the use of this object in the model.

Server

The **Server** is used to model a constrained resource such as a machine or a service operation that has a fixed location in the system. A Server could be used to model a check-in counter at an airport, an ATM machine, hotdog stand, barber chair, or a drill. The Server has an Input node where entities enter the Server and an Output node where they depart the Server. The commonly used properties for Server are summarized below:

-Process Logic	
Capacity Type	The capacity type specified as *Fixed* or *Work Schedule*.
Initial Capacity	The number of entities that can be simultaneously processed at this Server.
Ranking Rule	The static rule used to order the entities waiting for this Server.
Processing Time	The time required for this Server to process each entity.
-Buffer Capacity	
Input Buffer	The number of entities that can be held in the Server's Input Buffer
Output Buffer	The number of entities that can be held in the Server's Output Buffer
-General	
Name	The name of this object.
Description	A description of the use of this object in the model.

Entities enter a Server at the Input node, wait in an *Input Buffer* to begin processing, undergo some activity at the *Processing* station, and then move to the *Output Buffer* where they wait to depart the *Server* at the Output node. Hence at any given time all entities in the *Server* are either in the *Input Buffer*, the *Processing* station, or the *Output Buffer*. The *Capacity Type* specifies the method for controlling the capacity of the Server over time. By default the *Capacity Type* is *Fixed* and has an associated *Initial Capacity* property that defaults to 1 and specifies the number of entities that can be simultaneously processed at the Processing station. As we will see later the capacity can be assigned new values during the simulation to dynamically alter the value. The *Capacity Type* can also be specified as a *Work Schedule*; in this case the capacity automatically changes to follow a work schedule pattern that is defined in the Data window and specified as an associated property. The *Processing Time* property specifies the time required to perform the processing activity. This defaults to a random sample from a triangular distribution with a minimum of .1, mode of .2, and maximum of .3 minutes. The *Input Buffer* and *Output Buffer* properties specify the capacity for the input and output buffers; i.e. the maximum number of entities that can be in the Input Buffer or Output Buffer at any one time.

Note that by setting the *Initial Capacity* to a value greater than 1 a single Server object can be used to model a set of parallel servers. For example a barbershop with 5 chairs can be modeled with a single Server object with Initial Capacity of 5. This makes it very convenient to model multi-server systems in cases where each server has identical performance characteristics. Otherwise we can use a separate Server object for each of the parallel servers. The following figure depicts a Server with an Initial Capacity of 1, with two entities in the Input Buffer, one being processed, and one entity in the Output Buffer. Note that the entity in the Output Buffer may be held because a downstream Server or material handling device is blocking it from exiting the buffer.

The following depicts Server with an Initial Capacity of two. Note that in this case two entities can be processing at the same time.

By default the *Input Buffer* and *Output Buffer* on the Server has an infinite capacity. If this capacity is specified as a finite limit then the Server may block upstream entities, or may be blocked by downstream entities. For example if you place two Servers connected in tandem the second Server may block the first Server whenever the intervening Output and Input buffers are full. Note that the Input and Output Buffers are "inside" the Server and distinct from the Input and Output nodes that provide entry and exit points for the Server. Hence the buffer sizes are characteristics of the Server and not the associated Input/Output nodes.

Workstation

The **Workstation** is similar to Server. It has been simplified by limiting its design to model a single server while adding considerable flexibility for the behavior of that single server. In the Workstation the *Processing* activity is preceded by a Setup activity and followed by a Teardown activity. The Workstation has a single associated Input node and single associated Output node. The commonly used properties for the Workstation are summarized below:

- **Process Logic**
  - Capacity Type — The capacity type specified as *Fixed* or *Work Schedule*.
  - Ranking Rule — The static rule used to order the entities waiting for this Workstation.
  - Setup Time Type — The method used for specifying the setup time.
  - Setup Time — The time required to perform a setup for starting processing.
  - Processing Time — The time required to perform a processing activity.
  - Teardown Time — The time required to perform a tear down following processing.
- **Buffer Capacity**
  - Input Buffer — The number of entities that can be held in the Workstation's Input Buffer.
  - Output Buffer — The number of entities that can be held in the Workstation's Output Buffer.
- **Secondary Resources**
  - Secondary Resources — Opens an editor to define additional resources that are required during setup, processing and/or teardown.
- **General**
  - Name — The name of this object.
  - Description — A description of the use of this object in the model.

Entities enter the Workstation at the Input node, wait in an *Input Buffer* to begin a Setup activity, undergo Setup, Processing, and Teardown activities, and then move to the *Output Buffer* where they wait to depart the *Workstation* from the Output node.

The *Setup Time Type* defines the method that is used for determining the setup time and is specified as *Specific*, *Change Dependent*, or *Sequence Dependent*. The default type is *Specific* and has an associated property named *Setup Time* where the specific value for the setup time is entered. This value could be specified as a constant, an attribute of the entity, or a value defined for this entity in a data table. In the case of *Change Dependent* there are three associated properties; the *Operation Attribute* property specifies an attribute of the entity (e.g. size, color, part family, etc.) that controls the setup time, the *Setup Time If Same* property specifies the setup time if this attribute is the same as the previous entity, and the *Setup Time if Different* property specifies the setup time if is this attribute is different than the previous entity. The third option for *Setup Time Type* is *Sequence Dependent*, and is used to model situations where the setup times are sequence dependent based on a changeover matrix that is defined in the Data window. In this case two associated properties are specified; the first is the *Operation*

Attribute that specifies the attribute of the entity that is used in the changeover matrix, and the second is the *Changeover Matrix*.

Both the Server and Workstation support the concept of secondary resources. In learning mode the secondary resources are hidden on the Server, but they are partially exposed on the Workstation. A secondary resource is an object that is required as an additional resource to perform some activity; e.g. a special tool to machine a part, or an operator to perform a setup operation at a drill.

The *Secondary Resources* property on the Workstation has a button on the right with "...":

When clicked this opens up the *Secondary Resources Repeating Property Editor*. This is an editor (shown below) that is used to define one or more secondary resources that are required by the Workstation. The left side of the editor is a list of secondary resources that are added/deleted using the Add/Delete buttons, and the right side of the editor is used to view/change the individual properties of each secondary resource item.

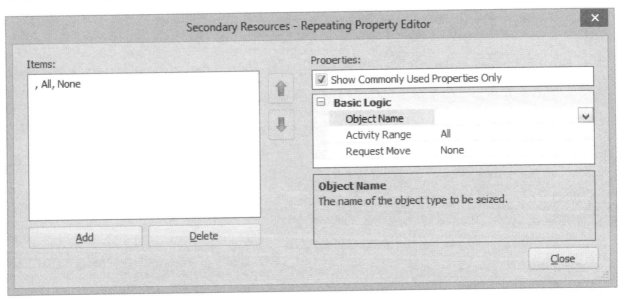

The first property is *Object Name* which specifies the resource object that must be seized. This can be a Resource object or Worker object from the Standard Library, or any other object that is defined to have resource behavior. The *Activity Range* property is specified as *All*, *Specific*, or *AllButSpecific* and is used to define the range of *Setup*, *Processing*, and *Teardown* activities where this secondary resource is required. This range defaults to *All* which causes the secondary resource to be utilized during all three activities. The *Specific* option is used to limit the utilization to a single activity which is specified as an associated property. The *AllButSpecific* is used to limit the utilization to two of the three activities, where the activity where it is not utilized is specified as an associated property.

If a moveable resource such as a Worker is specified for the secondary resource then the *Request Move* property can be used to specify if the resource object must first move to a new location before the activity can begin. If specified as *None* then the activity can begin as soon as the resource object is seized. If specified as *To Node* then the *Destination Node* property is required and specifies the node location to which the resource must move after being seized and before the activity can begin. The

Destination Node could be the Input or Output node associated with the Workstation, or more commonly a new free-standing node placed near the Workstation that represents the location where an operator stands when tending to the workstation.

The Workstation is considerably more flexible when all properties are displayed. Additional functionality includes a maximum makespan and consumption and production of a bill of materials. In addition an individual entity may represent a batch of work pieces, which is broken down into smaller batches for processing. Finally more complex logic can be used for selecting between resources in a list. All of this functionality has been removed here to focus on the core three-phase behavior of the Workstation.

Resource

The **Resource** is a generic object that can be seized and released by other objects. For example a Workstation representing a machine might require a special fixture as a secondary resource to process a specific part. This fixture could be modeled as a Resource that is seized and released as a secondary resource by the Workstation. In this case the drill is referred to as the primary resource, and the fixture is referred to as a secondary resource. Note that the Resource does not have an associated Input or Output node.

The commonly used Resource properties are:

-Resource Logic

Capacity Type	The capacity type specified as *Fixed* or *Work Schedule*.
Initial Capacity	The initial capacity of this resource.
Ranking Rule	The static rule used to order the objects waiting for this Resource.

-General

Name	The name of this object.
Description	A description of the use of this object in the model.

The Capacity Type defaults to Fixed but can also be changed to a Work Schedule to have a capacity that automatically varies over time. The capacity defines the number of units of this resource that can be in simultaneous use. For example a facility with 10 fixtures that are shared across a group of machines could be modeled as a single fixture resource with a capacity of 10. Each machine that requires the fixture would wait until one of the 10 units of the resource is available.

Note that the Resource object in the Standard Library is a fixed object that does not move during the simulation. The number of busy units of the resource object might change, but the location of all units of the resource remains fixed. Hence the Resource object is used to model resources where the location and movement of the resource is considered unimportant and therefore not explicitly modeled. In some cases the location and movement is important in capturing the behavior of the system. In this case the Worker object is a better choice for modeling the resource.

Worker

The **Worker** is a transporter object that can be seized and released by other objects, and also pickup and drop off entities and freely move from one location to another. The Worker might be used to model operators in a factory or doctors and nurses in a health delivery system. A Worker can be used as secondary resource for a Workstation (or Server in standard mode), but unlike the Resource object the Worker can travel from one location to another. When a Workstation requires the use of a secondary resource that is specified as a Worker the Worker may have to travel to a node location specified on the Workstation before processing can start.

The commonly used properties for the Worker are shown below:

-Resource Logic

Capacity Type	The capacity type specified as *Fixed* or *Work Schedule*.
Ranking Rule	The static rule used to order the objects waiting for this Worker.

-Travel Logic

Initial Desired Speed	The initial desired speed for workers of this type.

-Routing Logic

Initial Node (Home)	The initial node location for this worker at the beginning of the simulation run.

-General

Name	The name of this object.
Description	A description of the use of this object in the model.

The properties for the Worker are similar to the Resource except the resource capacity is automatically defined as 1, and the Worker has Travel and Routing Logic properties. The *Initial Desired Speed* is the initial desired travel speed for each of the workers of this type. The *Initial Node* is the initial location for the Worker.

Combiner and Separator

The **Combiner** is used to form a batch of Member entities and then combine the batch to a Parent entity. The Parent entity might represent a pallet or a tote, and the Member entities that items that are to be placed on/in the pallet/tote. The training mode properties for the Combiner are shown below:

-Matching Logic

Batch Quantity	The number of Member entities to be attached to the Parent entity.
Matching Rule	Determines how Member and Parent entities are grouped.

-General

Name	The name of this object.

Description	A description of the use of this object in the model.

The *Batch Quantity* defaults to 1 and specifies the number of Member entities that are to be attached to each Parent entity. The Matching Rule defaults to *Any Entity* which means that any waiting Member entities will form a group with any waiting Parent entity. The other rules, *Match Members*, and *Match Members and Parent* will expose additional properties to restrict matches to only entities with identical charactistics.

The Combiner has two input buffers that hold the Parent and Member entities. The Parent Input Buffer holds the Parent entities that are waiting to be combined with Member entities. The Member Input Buffer holds the Member entities that are waiting to be combined with a Parent entity. Once the member entities are attached to the Parent entity, the Parent entity is moved to the Output Buffer where it waits to depart the Combiner at the Output node.

The **Separator** is used to either separate the Member entities that have been attached to the Parent entity by a Combiner, or to make copies of an entity. The first mode is used to model situations where items are removed from container; e.g. work pieces removed from a pallet. The second mode is used to clone an entity; e.g. make a copy of a document and send the original to one location and the copy to a different location. The training mode properties for the Separator are:

-Separation Logic	
Separation Mode	Either *Split Batch* entities from Parent or *Make Copies*.
-General	
Name	The name of this object.
Description	A description of the use of this object in the model.

The *Separation Mode* property is specified as *Split Batch* or *Make Copies* and determines which of the two actions are to be performed by the Separator. In the case of *Split Batch* the Member entities are separated from the Parent entity. The Parent entity departs from the Parent Output node, and the separated Member entities depart from the Member Output node. The *Split Quantity* property determines how many items from the batch should be split, defaulting to all members. In the case of *Make Copies* a copy is made of the same type as the arriving entity (i.e. the arriving entity is cloned). The arriving entity departs from the Parent Output node, and the copy departs from the Member Output node. The *Copy Quantity* property determines how many copies to make and the Copy Entity Type property provides the option to create a new entity type (default is to create an identical entity).

Vehicle

The **Vehicle** is a transporter that can pickup and drop-off entities and freely move from one location to another. The Vehicle is used to model a wide range of transport type devices such as carts, trucks, cars, and buses.

The commonly used properties for the Vehicle are shown below:

- Transport Logic

   Initial Ride Capacity   The initial carrying capacity of this vehicle.

- Travel Logic

   Initial Desired Speed   The initial desired speed for vehicles of this type.

- Routing Logic

   Initial Priority   The initial priority assigned to this Vehicle.

   Initial Node (Home)   The initial node location of this vehicle at the start of the simulation run.

   Routing Type   Routing is *On Demand* or follows a *Fixed Route*.

- General

   Name   The name of this object.

   Description   A description of the use of this object in the model.

The first two properties are relatively simple; they define the number of entities that the vehicle can carry, and the initial desired travel speed for the vehicle. The *Routing Logic* properties include the *Initial Priority* that is assigned to the Vehicle, the *Initial Node* where the Vehicle is located, and the *Routing Type* that controls the movement for the Vehicle. The *Routing Type* property is the key property and specifies the routing behavior of the vehicle as one of two basic types. In the case of *On Demand* the vehicle responds to pickup request much like a taxi that is called for a pickup. In this request are handled in the order that they are received. In the case of *Fixed Route* the vehicle follows a defined route through a network of links and nodes. If the *Routing Type* is specified as *Fixed Route* the *Route Sequence* property is switched in to allow the user to specify a sequence table that defines the routing sequence of nodes to visit. Sequence tables are defined in the Data window of Simio and will be discussed later.

Vehicles (like Workers) typically move over a network of links and nodes. Links define pathways between nodes and nodes define the starting and ending points for links. As noted earlier many of the objects (e.g. Server) have associated nodes that define entry and exit points to the object and these nodes may also be part of a network. The following shows a small network with five nodes and six links. In this example three of the five nodes are free-standing and two of the nodes are associated with a Server object.

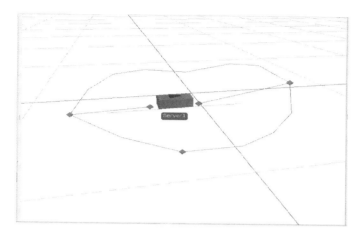

We will next look at the nodes and links in the Standard Library in more detail, beginning with the Basic Node.

Basic and Transfer Nodes

The **Basic Node** is a simple node that is used as the associated Input nodes for the Standard Library objects. It can also be placed as a free-standing node to provide a junction for one or more inbound or outbound links. The commonly used properties for the Basic Node are summarized below.

-Routing Logic	
Outbound Link Rule	The rule used by a traveler to select an outbound link from this node.
-General	
Name	The name of this object.
Description	A description of the use of this object in the model.

The Outbound Link Rule defines how a selection between multiple candidate outbound links is to be made and is specified as either *Shortest Path* or *By Link Weights*. If the traveling entity has a destination specified then the *Shortest Path* rule will select the link that is on the shortest path to that destination. If no destination is specified for the traveler then the *Shortest Path* rule defaults to the *By Link Weights* rule. The *By Link Weights* rule makes a probabilistic selection between the outbound links where the probability of selection is assigned in proportion to the link selection weight. For example if there are three outbound links with selection weights of 20, 30, and 50, then the links will be selected with a probability of .2, .3, and .5, respectively. Link selection weights are specified directly on the link and can be either simple numbers, or conditions. As we will see later link selection weight conditions provides a very simple yet powerful way to specify complex routing in a network.

The **Transfer Node** is a slightly more complex node that is used as Output nodes in the Standard Library objects. The Transfer Node can also be used as a free-standing node in a network of links and nodes. The commonly used properties for the Transfer Node are summarized below.

-Routing Logic		
	Outbound Link Rule	The rule used by a traveler to select an outbound link from this node.
	Entity Destination Type	A rule used for setting the destination node for entities departing this node.
-Transport Logic		
	Ride On Transporter	Specifies if entities departing this node must ride on a transporter.
-General		
	Name	The name of this object.
	Description	A description of the use of this object in the model.

The first property is the *Outbound Link Rule* and is the same as the Basic Node. The two additional properties are *Entity Destination Type* and *Ride On Transporter*. Each of these properties has additional properties that may be switched in depending on their value.

The *Entity Destination Type* is used to set a destination node for the departing entity and is specified as *Continue, Specific, Select from List*, or *By Sequence*. In the case of *Continue* the entity departs the node without resetting its destination node; i.e. it keeps its current destination node value (if any). In the case of *Specific* a *Node Name* property is switched in that is used to specify a specific node to set as the new destination for this entity. In the case of a *Select from List* a *Node List Name* property is switched in to specify the name of the node list from which a selection is to be made. Here this selection is made in preferred order from the available nodes in the list – where a node is defined as available if it's not currently blocking based on the remaining capacity of the input station at its associated object. For example when selecting from the Input nodes from a set of Servers, a Server Input node is blocking if its *Input Buffer* is at capacity, otherwise it is available. Note that a Server Input node is available even when the Server is busy as long as it has space available in its Input Buffer. If all Servers are blocking (all Input Buffers full) the entity will wait at the Transfer Node until one of the downstream Servers becomes available.

To use the *Select from List* option it is first necessary to create a list of nodes. The easiest way to do this is to use the mouse control-click to create an extended selection set of nodes, and then from the right click menu on any of the selected nodes click on *Add to Node List*.

The final option for the *Entity Destination Type* is *By Sequence*. This option specifies that the entity is to assign the next destination in the node routing sequence that the entity has been assigned to follow. As we will discuss later this sequence table is created in the Data window and assigned to the entity using the *Initial Sequence* property on the ModelEntity.

The *Ride On Transporter* flag is used to specify if entities departing this node must do so by riding on a transporter. If specified as True then a Transporter Type property is switched in to specify if a *Specific* transporter is required, or if the transporter is to be selected *From List*. In the case of *Specific* the *Transporter Name* property is switched in to select the name of the specific transporter. In the case of *From List* the Transporter List Name property is switched in to select the name of the transporter list from which a transporter selection is to be made. Here this selection is made based on the closest transporter, with ties broken by list order. A transporter list is created the same way as a node list – by creating an extended selection set of transporters and then selecting *Add to Transporter List* from the right click menu.

Connector, Path and TimePath

The **Connector** is a type of link that is used to define a direct, zero-time travel from one node location to another. It is used to model situations when an entity moves from one location to another in a negligible time. The commonly used properties for the Connector are:

-Routing Logic

Selection Weight	The selection weight for this outbound link when using a By Link Weight rule.

-General

Name	The name of this object.
Description	A description of the use of this object in the model.

When running a model the entity traversal of an entity on a Connector is not animated since the entity traverses the Connector in zero time. To see movement on a link use the Path, Time Path, or Conveyor.

The **Path** is a link that is used to model a pathway over which entities travel at their own independent speed. The travel time through the Path is determined by the length of the Path and the speed of the entity. A Path is used to model a wide range of pathways such as a road, sidewalk, aisle, or railroad track. The commonly used properties for the Path are summarized below:

-Travel Logic

Type	The type of travel specified as *Unidirectional* or *Bidirectional*.
Allow Passing	Indicates whether passing is allowed on his path.

-Routing Logic

Selection Weight	The selection weight for this outbound link when using a By Link Weight rule.

-State Assignments   Repeat group to allow states to be changed on entry to this link

State Variable Name	The name of the state variable to change.
New Value	The new value of the state variable.

-General

Name	The name of this object.
Description	A description of the use of this object in the model.

The *Travel Logic* includes the *Type* of travel and an *Allow Passing* flag. If the travel *Type* is *Unidirectional* then all traffic must travel in the forward direction that the Path was drawn. If the travel Type is

Bidirectional then the traffic direction can switch back and forth between a forward and reverse direction, however only on direction of traffic flow can be active at a time. Hence the traffic flow direction can only switch once the Path is clear of all traffic. The *Allow Passing* property defaults to *True* and specifies whether entities are allowed to pass each other on the Path. Note that when *Allow Passing* is *True* and multiple entities are waiting at the end of the Path the animation will make it appear as only one entity is at the end of the Path. The reason for this is that the passing entities overlap each other.

The **TimePath** is a link that is used to model a pathway over which entities travel in a specified time. The length of the TimePath has no impact on the travel time; the speed of the entity is automatically adjusted to travel the TimePath in the specified time. The commonly used properties for the TimePath are summarized below:

-Travel Logic	
Type	The type of travel specified as *Unidirectional* or *Bidirectional*.
Travel Time	The time required to travel the length of the path.
-Routing Logic	
Selection Weight	The selection weight for this outbound link when using a By Link Weight rule.
-State Assignments	Repeat group to allow states to be changed on entry to this link
State Variable Name	The name of the state variable to change.
New Value	The new value of the state variable.
-General	
Name	The name of this object.
Description	A description of the use of this object in the model.

Conveyor

The **Conveyor** is a link used to model a device such as a belt, rollers, buckets, etc., is used to transport the entities along a pathway. The speed of the Conveyor controls the speed of the entities that ride on the device. The commonly used properties for the Conveyor are summarized below:

-Travel Logic	
Initial Desired Speed	The initial desired speed of this conveyor object.
Entity Alignment	Entity spacing specified as *Any Location* or *Cell Location*.
Accumulating	Specifies whether this conveyor is accumulating.
-Routing Logic	

Selection Weight	The selection weight for this outbound link when using a By Link Weight rule.
-General	
Name	The name of this object.
Description	A description of the use of this object in the model.

The spacing of entities on the Conveyor is determined by the Entity Alignment property. If specified as *Any Location* the entity can merge onto the conveyor at any location. If specified as *Cell Location* the entity can only merged at fixed cell locations along the conveyor, where the spacing is specified by the associated *Cell Size*. When an entity is delayed at the end of the Conveyor and *Accumulating* is specified as *True* the entity "slips" on the Conveyor and allows the Conveyor to continue to move other entities; otherwise it stops the Conveyor until the entity is removed from the end of the Conveyor. Note that this slippage action may cause entities to accumulate at the end of the Conveyor.

ModelEntity*

An entity typically represents a part, person, or other object that is dynamically created, flows through the system, and then leaves the system. You will find one ModelEntity automatically created within a new Project Library. The **ModelEntity** is the default entity definition in your project. A user does not need to drag it into the Facility window unless you want to customize its properties or animation. A ModelEntity called DefaultEntity is already part of the Project and can therefore be created by a Source object or with the Create Step without having an instance in the Facility window. Additional instances of ModelEntity can also be dragged into the model. The commonly used properties for ModelEntity are summarized below:

-Travel Logic	
Initial Desired Speed	The initial desired speed of an entity.
-Routing Logic	
Initial Priority	The initial priority, often used for ranking and selection.
...Initial Sequence	The initial sequence used when routing By Sequence.
-General	
Name	The name of this object.
Description	A description of the use of this object in the model.

*ModelEntity is found in your Project Library, not the Standard Library

Chapter 4: Our First Model

Overview

The object libraries provide definitions for objects that can be placed into the facility model. There are several different ways to select and place objects into the facility model. For all objects except for links you can use any of the following methods:

1. Click on the object definition in the library (e.g. Source, Server, or Sink), and then click at the location in the Facility window where you would like to place the object.
2. Drag the object from the library to a location in the Facility window.
3. Double-click on the object definition in the library, click in each location in the Facility window where you would like to place the object, and then terminate multi-place mode by right clicking anywhere in the model.

Once you have placed an object in the facility you can move it by clicking on it and dragging it. You can change the default name by double-clicking on the name or by right-clicking on the object. You can resize objects (except for nodes) by dragging the resize handles that appear once the object has been selected.

In most cases placing an object will instantiate associated nodes as well as the main object. For example if you place a Server object named Server1 in your facility you are actually placing three separate but related objects. The main (center) object is the server itself and has a set properties related to the service operation. The small diamond object on the left is the associated input node named *Input@Server1* (a BasicNode) and you can move it independently of the server, and it also has its own set of properties which you may edit. The small diagonal object on the right is the output node named *Output@Server1* (a TransferNode) and you can select and move it independently and also edit its properties. If you move the main Server object it will also automatically drag with it the associated input and output node objects.

Links are special because they provide a pathway between two nodes, and hence must always start and end at a node. Links are not "placed" but are instead drawn as a polyline between two nodes. You can use either of the following methods to draw links.

1. Click on a link object in the library (e.g. Connector, Path, TimePath, Conveyor), and then click on the starting node for the link pathway, click at each intermediate vertex that you would like to add, and then click on the ending node for the pathway. You can abort link drawing by right-clicking before completing the link.
2. Double-click on a link object in the library, add a link instance as described above, and then repeat multiple times. Terminate multi-link draw mode by right clicking anywhere in the model.

Once you have drawn a link you can move any of its vertices by clicking and dragging the vertex.

Manipulating Facility Views

By default Simio provides a top-down 2-D view of the facility model. This is often a very convenient view for creating and editing your model. However facility models in Simio are defined in 3D. You can switch between the 2D and 3D views of the facility using the View section of the View ribbon, or by using the 2 and 3 keys on the keyboard. The View section also lets you set the facility model to auto-rotate in 3D – click anywhere in the facility to stop auto-rotation. You can also change the background color of the 3D space. Click anywhere in the facility and drag to pan the view. Press the right mouse button and move left and right to rotate the view, or up and down to zoom in and out. You can also zoom using the mouse wheel, or change the angle of the camera with the floor by holding the control key and moving the mouse wheel.

You can also position the camera to make it look like you are walking on the floor by pressing the W key. The forward, left, right, back keys then move you across the floor in the direction of the key.

During the running simulation you can have the camera movement controlled by a moving entity. To do so pause the running model and right click on a moving entity and select *Track with Camera*. From the View ribbon you can then select between camera placement options that include *Move camera with object*, *Watch from a distance*, *Follow behind*, and *Lead in front*.

You can use the Named Views section of the View ribbon to create and switch to named views into the facility model – such as the assembly area or shipping. This is a useful feature for navigating through very large models. Camera Sequences may also be defined to switch between various views in a specific sequence, staying at each view for a designated time period. The video section of the View ribbon also allows for video recording of the simulation animation.

The Visibility ribbon lets you selectively hide/show different aspects of the facility model – e.g. you can hide all nodes and links.

Editing Object Properties

Whenever you select an object in the facility (by clicking on it) the properties of the selected object are displayed for editing in the Property window in the Browse panel on the right. For example, if you select a Server object in the facility the properties for the Server will be displayed in the Property window. The properties are organized into categories that can be expanded and collapsed. In the case of the Server in the Standard Library the Process Logic category is initially expanded, and all others are initially collapsed. Clicking on +/- will expand and collapse a category. Whenever you select a property in the property grid a description of the property appears at the bottom of the grid.

The properties for an object are defined by the designer of that object. The properties may be different types such as strings, numbers, selections from a list, and expressions. For example the Ranking Rule for the Server is selected from a drop down list and the Processing Time is specified as an expression.

For expression fields (e.g. Processing Time) Simio provides an expression builder to simplify the process of entering complex expressions. When you click in an expression field a small button with a down arrow appears on the right. Clicking on this button opens the expression builder. The expression builder is very similar to IntelliSense® in the Microsoft family of products and tries to find matching names or keywords as you type. You can use math operators +, −, *, /, and ^ to form expressions. You can use parenthesis as needed to control the order of calculation (e.g. *2*(3+4)^(4/2)*). Logical expressions can be expressed using <, <=, >, >=, ==, !=, &&, and || (the last four are logical Equals, Not-equal, And, and Or). Logical expressions return a number value of 1 when true, and 0 when false. For example *10*(A > 2)* returns a value of 10 if *A* is greater than 2, and otherwise returns 0. Arrays (up to 10 dimensions) are 1-based and are indexed with square brackets (e.g. *A[2,3,1]* indexes into the three dimensional state array named *A*). Model properties can be referenced by property name. For example if *Time1* and *Time2* are properties that you have added to your model then you can enter an expression such as *(Time1 + Time2)/2*.

A common use of the expression builder is to enter random distributions. These are specified in Simio in the format Random.DistributionName(Parameter1, Parameter2, …, ParameterN), where the number and meaning of the parameters are distribution dependent. For example Random.Uniform(2,4) will return a uniform random sample between 2 and 4. To enter this in the expression builder begin typing "Random". As you type the highlight in the drop list will jump to the word Random. Typing a period will then complete the word and display all possible distribution names. Typing a "U" will then jump to *Uniform(min, max)*. Pressing enter or tab will then add this to the expression. You can then type in the numeric values to replace the parameter names min and max. Note that highlighting a distribution name in the list will automatically bring up a description of that distribution.

Math functions such as Sin, Cos, Log, etc., are accessed in a similar way using the keyword Math. Begin typing "Math" and enter a period to complete the word and provide a list of all available math functions. Highlighting a math function in the list will automatically bring up a description of the function.

Object functions may also be referenced by function name. A function is a value that is internally maintained or computed by Simio and can be used in an expression but cannot be assigned by the user. For example all objects have function named *ID* that returns a unique integer for all active objects in the model. Note that in the case of dynamic entities the *ID* numbers are reused to avoid generating very large numbers for an ID.

Some other features of Simio's expression building interface are illustrated below:

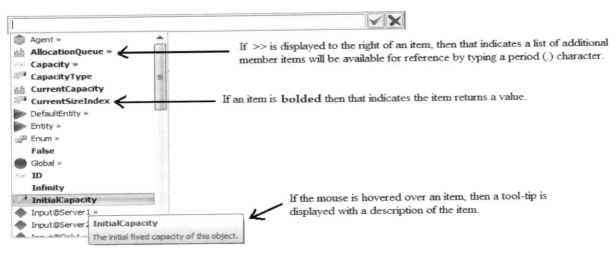

If you type an invalid entry into a property field the property name turns a salmon color and the error window will automatically open at the bottom. If you double-click on the error in the error window it will then automatically take you to the property field where the errors exist. Once you correct the error the error window will automatically close.

A Simple Flow Line Model

We will now use the Standard Object Library to build a model of a simple flow line in which entities arrive to the system, are processed through a sequence of two servers, and then depart the system.

Introduction to Simio

We will model the arrival process of customers using the Source object, each server using the Server object, and departure process using the Sink object. We will use the Path object to model the pathways between these objects.

The mechanics of building this model is very simple. Open Simio to the empty Facility window – if you already have a model loaded, select **File** -> **New** to create a new blank project. Drag and place the Source, two Servers, and a Sink. Now double click on the Path object to enter path drawing mode and connect the output node of the source to the input node of Server 1, connect the output node of Server 1 to the input node Server 2, and finally connect the output node of Server 2 to the input node of the Sink. Click anywhere in the drawing space to terminate path drawing mode. Now click on the **Run** button on the left side of the Run ribbon and you will see your model start executing as shown below.

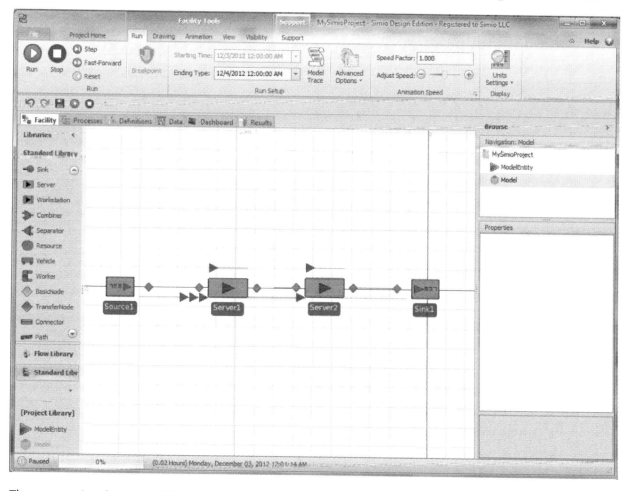

The green triangles are entities that represent the customers moving through the system. Note that each server has three green lines representing animations of the queues owned by the server. The animated queue on the left of the server is the input buffer where customers wait their turn to be processed, the animated queue above the server is the processing buffer where entities sit while being serviced, and the animated queue on the right is the output buffer where entities wait to leave the server. You can move and resize these queue animations. Note that these are animation-only constructs -- lengthening the animation queue changes the number of customers that can be animated in the queue, but not the actual number of customers that can be in the logical queue. The latter is changed by selecting the server and changing the buffer capacities in the Properties window. Also note that the actual number in the queue may exceed the number shown in the animated queue and it may be necessary to lengthen the animation queue to see all of the entities that are in the queue.

Note that you can pause and restart the simulation, or stop the execution of the model. You can also step an entity at a time, fast-forward the run without the animation, or reset the model back to its starting conditions. You can perform graphical edits such as moving a Server, adding a new path, etc., while the model is running, but you must stop the simulation to change any of the object properties.

At this point the model is running with all default property values. We will now modify some of the object properties so that we can see the impact of changing the waiting space between *Server1* and *Server2* on the upstream blocking of *Server1* by *Server2*. Click Stop to stop the model from running and click on *Server1* to display its properties in the Properties window. Expand the *Buffer Capacity* category by clicking on the + sign, and set the *Output Buffer* capacity to 0 to eliminate any waiting space in the output buffer. Next click on *Server2*, expand the *Buffer Capacity* category, and set the *Input Buffer* capacity to 0 to eliminate any waiting space in the input buffer. With these buffers now set to zero, all entities between *Server1* and *Server2* are forced to be on the path connecting the two servers. Now click on this path and set the *Allow Passing* property to *False*, which will force the entities to back up behind each other on the path. If this path becomes backed up all the way to *Server1*, this server will be blocked from processing the next customer until space becomes available on the path.

Next we will add a status pie chart to each server so that we can animate the resource status of each server as the model runs. Click on *Server1* which will also automatically select the Symbols ribbon. Click on the small down arrow to the right of Status Pie, and then select *ResourceState* from the drop list. Click in the Facility window to add the vertex for the top left of the pie, and then click again to add the vertex for the bottom right. This will add a pie chart that will animate the resource state for *Server1*. You can drag and resize this as needed. Repeat these same steps for *Server2*.

We will now improve our animation by replacing the green triangle with a 3D graphic of a person. Drag a *ModelEntity* from the Project Library on the lower left panel and place it in the facility window. Note that the *ModelEntity* is in the Project Library because it exists as a model (object) in our project. You will see a green triangle with the name *DefaultEntity*. Note that if you select the Source object the *Entity Type* property specifies this *DefaultEntity* as the type of entity to create. The *DefaultEntity* is special in that Simio automatically creates this entity for you without you having to actually place it in your facility. This is done so that simple models will run without having to place an entity in the facility and specify the entity type on the source. This specially created entity becomes the first one that is placed in the facility and any additional entities that you place will be created only when you actually place them in the facility. Now click twice on the entity name (or just hit F2) and rename the *DefaultEntity* to *Customer* – note the name is automatically changed for the *Entity Type* property on the *Source* as well. Now that we have placed the default entity (now named *Customer*) we can change its graphical appearance. Note that by selecting the *Customer* you can resize it, or select a color or texture from the ribbon and then click on the entity symbol to apply the selection. To change the symbol click on the down arrow in the symbol pane to drop the list of symbols, scroll through the list and select a symbol of a person. The green triangle will now be replaced by that symbol. You may wish to switch to 3D (by pressing the 3 key) to view and resize the symbol within the facility.

With these changes our running animation now appears as follows. Since *Server1* is blocked by *Server2* whenever the pathway between them is full you can alter the amount of blocking that occurs for *Server1* by moving the servers to change the length of the pathway between them.

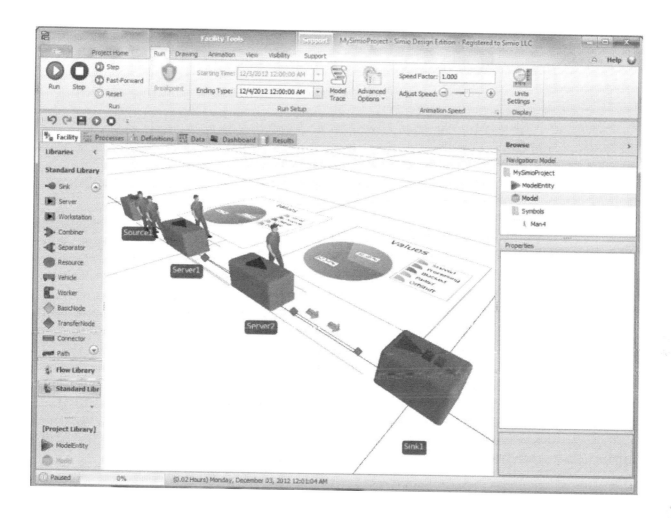

Defining Model Properties and Experiments

There are two basic modes for executing models in Simio. The first mode is the interactive mode that we have been using up to this point. In this mode you can watch the animated model execute and view dynamic charts and plots that summarize the system behavior. This is useful for building and verifying/validating the model as well as getting general insight into how the system will perform. Once the model has been validated the next step is typically to define specific scenarios to test with the model. In this case we have no interest in the animation and we would like to replicate each scenario to account for the underlying variability in the system and to reach statistically valid conclusions from the model. We will now turn our attention to the experimentation mode of Simio.

In the experimentation mode we define one or more properties on the model that we can change to see the impact on the system performance. These properties might be used to vary things like conveyor speeds, the number of operators available, or the decision rule for selecting the next customer to process. These model properties are then referenced by one or more objects in the model.

Continuing with our simple example let's assume that we have three grades of equipment that we could use, each with a different processing time. We will add a model property that we can reference and use to control the processing time on *Server2*. We will then run some experiments to see the impact of different processing times by changing the value of this model property. One way to manually add this property to the model is to select the Definitions tab for the model, select the Properties panel on the left, click in the ribbon on the down arrow for the Standard Property, and select Expression. We could then return to the Facility window and select this newly created property for the *Processing Time* for *Server2*. However a short cut method to accomplish these same steps is to right click on the *Processing Time* property in the Property window for *Server2*, and select **Set Referenced Property > Create New Referenced Property**. This will create a new reference property with the default name *ProcessingTime*, and also specify this as the value for the *Processing Time* property. Hence we create the new property and set the reference to it in a single step.

Now that we have our new referenced property we will define an experiment to see the impact of varying this property. Add a new experiment by right-clicking on **Model** in the navigation tree, and selecting **New Experiment**. This will create a new experiment with default name *Experiment1* and also select this experiment as the active project component. You can easily return to your model by clicking on Model in the navigation tree. You can create as many experiments as you like for your model. For example you might have one experiment to evaluate processing times, and another experiment to evaluate buffer sizes or staffing levels.

Our experiment shows a table, where each row of the table is a **scenario** to be executed. Note that each scenario has a check box (enabled/disabled), name, status (initially Idle), and replications required and completed, and a column for each referenced property on the model. The properties which are referenced by the model are referred to as **Controls** since they control the inputs to the model for the scenario being run. In this example we have a single referenced property named *ProcessingTime* that is used to control the processing time for *Server2*.

We can also add one or more **Responses** or **Constraints** to our experiment. A Response can be any valid expression and is typically a key performance indicator (KPI) that is of particular interest to us in comparing the scenarios. Although the Pivot Grid and Reports will provide detailed results for the model, we typically have a few key parameters that we would like to display directly in the experiment table. We can also select a specific response to be the **Primary Response** for the experiment that is displayed in the Response Chart for the experiment. A Constraint is an expression that must be satisfied for the scenario to be valid. Both Responses and Constraints may be used by custom add-ins for defining and evaluating the results from an experiment. For example the OptQuest add-in may be used to perform an automatic search to optimize the Primary Response, subject to any constraints that have been defined.

Click on *Add Response* in the ribbon to add a new response to your experiment. In the property grid change the response Name to *TimeInSystem*, enter the Expression as *Sink1.TimeInSystem.Average*, Unit Type as Time, and select the Objective as Minimize.

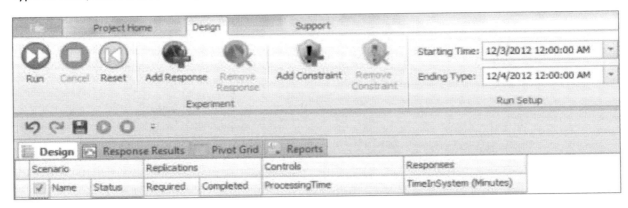

We will now define and run three scenarios which we will name *Small*, *Medium*, and *Large*. Enter these names in the *Name* column of the table and change the number of required replications to 50. In the *Process Time* column change the parameters of the triangular distribution for the *Medium* scenario to 0.1, 0.23, 0.33, and in the *Large* scenario change these values to 0.1, 0.26, 0.36.

Now click on the Run button to initiate the batch running of these scenarios. When you do so you will see the status for one or more scenarios Running, and the remaining ones listed at Pending. As each scenario completes its 50 replications it status will change from Running to Completed. If you are running on a multi-core processor Simio will automatically assign replications to be executed across each core of the computer. For example if you have a quad-core processor Simio will run four replications at a time.

Interpreting the Results

Your specific results for this example will depend upon the length for your paths and size of your customers. However the results should look similar to the following:

Scenario		Replications		Controls	Responses
Name	Status	Required	Completed	ProcessingTime	TimeInSystem (Minutes)
001	Completed	50	50 of 50	Random.Triangular(.1,.2,.3)	0.716348
002	Completed	50	50 of 50	Random.Triangular(.1,.23,.33)	0.826378
003	Completed	50	50 of 50	Random.Triangular(.1,.26,.36)	1.02619

Note that in this case the *Medium* and *Large* responses for *TimeInSystem* are grayed out. This is the result of clicking on the Subset Selection button in the ribbon which causes Simio to automatically separate the scenarios into two sets for each response: those that might be the best (shown in solid), and those that can be discarded as not the best (shown grayed-out). This indicates that the *Small* scenario can be statistically selected as the single "best" scenario based on minimizing this response. If the response has more than one member of the "could be best" you can narrow the selection down to a single scenario using the *Select Best Scenario using KN* add-in. This add-in employs the sequential selection method by Kim and Nelson (S. Kim and B. L. Nelson, "A Fully Sequential Procedure for Indifference-Zone Selection in Simulation," *ACM Transactions on Modeling and Computer Simulation* **11** (2001), 251-273) to automatically make the required additional runs to narrow the selection down to a single best scenario.

You can also gain additional insight into your responses by viewing the **Response Results** for the primary response. Click on the Response Results tab which brings up the following panel and ribbon.

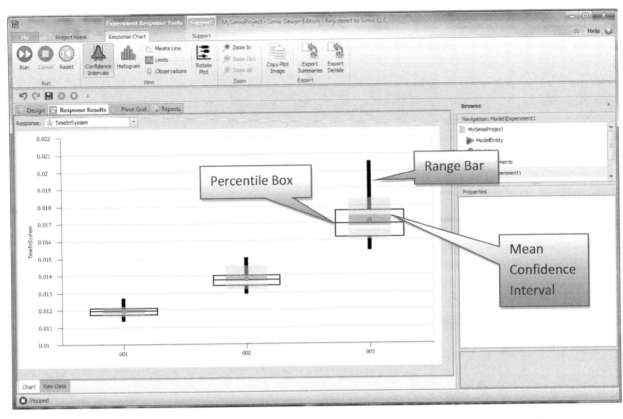

The Response Results shows a SMORE chart (an enhanced version of Nelson's MORE chart) that graphically displays a number of different summary statistics for each scenario. The black rectangular box shows the region containing observations between a lower and upper percentile point (by default 25% and 75%). The black bar shows the range of all values. The tan box is the confidence interval on the mean, and the blue boxes are confidence intervals on the lower and upper percentile points. The dark tan area is where the confidence intervals overlap. You can also show individual observations, a histogram, limits, or a line plot by toggling these on in the ribbon. You can also rotate the entire chart to view it either vertically or horizontal.

Whenever you save your project the results for each experiment are also saved. You can discard the results by clicking on the Reset button in the Design ribbon. After running an experiment you can also view the results and add additional replications (by changing the Replications Required) to reduce the half-width of the confidence intervals.

You can also view the detailed results for a wide range of statistics that are recorded by default during the run under either the Pivot Grid or the Reports tab. The Pivot Grid is convenient for quickly drilling into results of interest by letting you interactively analyze the results in the form of a dynamic rotatable table. You can drag and drop columns in the table to rotate or "pivot" the data. You can also filter and sort the information that is shown. In just a few clicks you can create a custom report and save it for later reuse. You can also export the results to external programs such as Excel. The Reports tab displays a more traditionally formatted report and provides flexible visual formatting for inclusion on web pages or in printed documents and is useful in presenting results to others.

The results in the Pivot Grid and the Reports show the average, minimum, maximum, and half width for each of the variables that have been recorded during the simulation. By default the objects in the Simio

Standard Library automatically record a wide range of statistics. As you will see later you can selectively control the statistics that are recorded.

Summary

This chapter has introduced the basic modeling concepts of Simio. We have introduced the concept of a Simio project as a collection of models and associated experiments. We have also introduced the basic concept of an object and how it is used as a building block for rapidly building and running 3D models in both an interactive and experimentation mode. In subsequence chapters we will expand on this basic knowledge to fully explore the modeling power of Simio.

Chapter 5: Network Travel

Overview

In Chapters 2 and 4 we introduced the basic concepts for building and running object-based models in Simio. In our first example we modeled a single line of flow with entities arriving from the Source, proceeding through the two in-line Servers, and then departing at the Sink. In this chapter we will look at Simio features for controlling entity and transporter movements. Note that transporters are a special case of entities: they are entities that can also pick up, carry, and drop off other entities. Hence our description of entity movements will also apply to transporter movements.

Entities may travel in either free space or over a network of links and nodes. In free space travel an entity may move without the aid of other objects by setting its direction, speed, and acceleration. In this case the movement is only governed by the entity and it can freely and independently move through 3D space. In the case of network travel the entity moves over a constrained set of links and nodes. In this case the movement is governed by both the entity and the corresponding links and nodes. Note that different types of links and nodes can produce different types of travel behavior based on the internal models for those links and nodes. For example one link might impose conveyor-like behavior, and another link might allow for independent movement and passing. In this chapter we will focus on network travel.

Entities and Attributes

Entities are dynamic objects that can be created and destroyed, travel over networks, and enter and depart fixed objects thru their associated node objects. The dynamic entity objects are created from an entity instance. You typically place one or more entity instances in your model by dragging them in from your Project library or another custom library. The entity instance has a stationary location and is used as a "template" for generating dynamic entities that move through the model. Each of the dynamic entities may have both properties and states (collectively referred to as attributes). The properties are defined by the entity instance and are unchangeable during the run and shared across all the dynamic entities. However each dynamic entity will have its own values for its states which may change as the model runs.

For example, assume we have an entity definition named *Person* that has properties named *Sex* and *HairColor*, and states named *Income* and *MaritalStatus*. Assume that we place two instances of *Person*; one we name *CustomerTypeA* and the other we name *CustomerTypeB*. We can specify the values for *Sex* and *HairColor* for either *CustomerTypeA* or *CustomerTypeB*. All of the dynamic entities that are generated for *CustomerTypeA* will all have the same value for these properties, and these values cannot change during the run. In the same way all of the dynamic entities that are generated for *CustomerTypeB* will have the same non-changeable values for *Sex* and *HairColor* as defined by the statically placed entity instance for *CustomerTypeB*. However each dynamically created *Person* (from either *CustomerTypeA* or *CustomerTypeB*) will have its own unique value for *Income* and *MaritalStatus* and these values may be changed during the run.

A **state** may be added to an entity by selecting the entity model (e.g., ModelEntity) in the navigation window, selecting the States panel in the Definitions window, and then clicking on one of the state icons in the States ribbon. Note that a state added to your main model can be thought of as a global variable, while a state added to an entity model or other object may be thought of as an attribute of that entity or object.

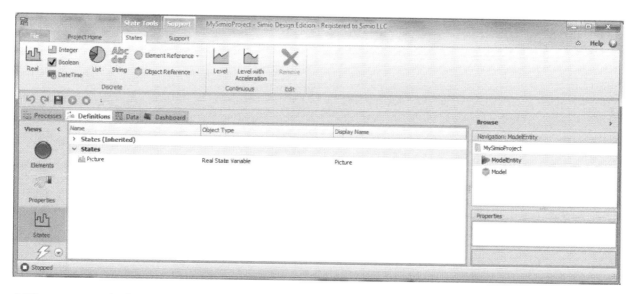

A **Discrete state** is the most commonly used state variable. Discrete states can be of type *Real, Integer, Boolean, DateTime, List, String, Element Reference* or *Object Reference*. By default a discrete state variable is a simple *Scalar*, but it can also be dimensioned as a *Vector*, a *Matrix*, or an array up to 10 dimensions. In the latter case you enter a number for *Dimensions* instead of selecting from the drop list. In the case of a vector or matrix you can also both dimension and initialize a state variable using the values in a data table by selecting the Dimensions property as [Table]. In this case the number of rows and columns for the matrix are set to the number of rows and columns in the table, and the matrix is initialized to the table values. This is a very convenient way to both dimension and initialize arrays using external data.

A discrete state of type *List* is a state variable that can be assigned a value from a list (list are defined using the List panel in the Definitions Window). For example a list named *Color* with *members Red, Green, Blue* could be used to define a list state name *ColorValue*. The numeric value of a list is referenced using the keyword *List*. Hence assigning *ColorValue* to *List.Color.Blue* would set *ColorValue* to the color list member *Blue*. All discrete state values return a numeric and in the case of a list state this is the zero-based index into the list, or in this case a value 2 since *Blue* is a zero-based index 2.

Discrete states of type *Element Reference* or *Object Reference* may store information related to a particular element, such as a Tally Statistic, or an object, such as an entity, node or transporter. These references may be changed using an Assign step, similar to assigning a real or integer type state.

A **Continuous state** is a state whose value may change continuously and automatically and may be of type *Level or Level with Acceleration*. A **Level state** is described by both a rate and level. For example if you create a level state named *Tank*, you can reference the level value by *Tank*, and the corresponding rate value by *Tank.Rate*. Note that the *Tank* value automatically updates based on the current value assigned to *Tank.Rate*. A **Level with Acceleration** is a state that has a level and is described by both a rate and acceleration.

Note that if you are adding state variables to one more entity models, and then creating objects for those entities, you must reference the states of the entity by prefixing the state names with the entity model name. For example if you create an entity model named Part, and then give it discrete states X and Y, you reference these states as Part.X and Part.Y.

Networks

A network is a collection of one or more links over which entities travel. In Simio you can define as many networks as you need, and a link can be a member of multiple networks. This latter point is important for modeling situations where different entities travel on their own networks, but share a common pathway: e.g. workers and forklift trucks sharing a common aisle. For network travel an entity must be assigned to a specific network over which they are permitted to travel. The collection of all links is automatically assigned to a special network called the Global network. By default all entities are assigned to travel on the Global network (i.e. they can travel on any link). The entity may change its travel network at any time during the running of the simulation.

Consider the following graph of nodes and links, in which links A, B, C, D, and F are in network 1, and links E, F, and G are in network 2. Note that in this case we have link F common to both networks. The red and blue links correspond to network 1, and the green and blue links correspond to network 2. Note that the arrows in the graph denote unidirectional links. We will discuss bi-directional links later in this chapter.

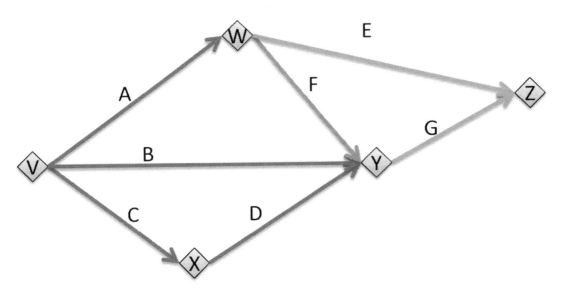

Network 1: A, B, C, D, F

Network 2: E, F, G

Global Network: A, B, C, D, E, F, G

By assigning entities to networks we can control where in the system they can travel. For example we might assign a person entity to the Global network, an AGV to network 1, and a forklift truck to network 2. In this case people, AGV's, and forklift trucks could all travel from W to Y along link F. The initial network is assigned to an entity by clicking on the entity instance in the facility window (you must first place the entity instance by dragging it from the Project library) and then setting the *Initial Network* property.

In Simio a link may be added to an existing or new network or removed from an existing network by right clicking on the link. The links that belong to a specific network can be highlighted using the View

Networks drop list on the Visibility ribbon. If multiple networks are selected for highlighting, then either the Union (links that are a member in one or more of the selected networks) or Intersection (links that are a member of all selected networks) can be highlighted.

Node Routing Logic

Whenever an entity departs a node with more than one departing link in its network then the decision of which link to travel along is based on the node routing logic. For example an entity traveling on network 1 and departing node V will have 3 links to choose from based on its routing logic. However an entity traveling on network 1 and departing node W has only one option: i.e. link F.

In large and complex models we often have alternate pathways through the system and must specify routing logic for controlling the movement of entities through the system. In the case of the Standard Library the routing logic for selecting between multiple candidate links is specified by an outbound link rule and outbound link preference. The outbound link rule specifies how the link is to be selected, and the outbound link preference specifies if all links are to be considered (*Any*), or only those that are currently available (*Available*). An available link is one that can immediately be entered. The outbound link rule is specified as either *Shortest Path* or *By Link Weight*.

In the case of shortest path the entity will select the path that is along the shortest path to its destination (note that we will discuss the options for setting the destination for an entity in a moment). Hence a person travelling on the Global network from V with its destination set to Z would select link B at node V (assuming B is available or the *Any* option is selected). If the outbound link rule is specified as shortest path but the destination is not assigned, then the link is selected by link weights.

When outbound link rule is specified as By Link Weight (or no destination has been set), then the link is selected randomly using the weights specified for each candidate link. The probability of a given link being selected is equal to its weight divided by the sum of the weights for the candidate links. All links have a selection weight property that can be specified as a number or any valid expression. In the simple case we can use these link weights like percentages or probabilities. For example if one link has a weight of 80 (or .8) and the other a weight of 20 (or .2), then the links will be selected randomly with an 80/20 split.

In Simio a logical expression (e.g. $X > 2$) will return a 1 if true, and 0 if false. This makes it very convenient to use logical expressions for specifying link selection weights. Note that in this case the weights will dynamically change during the simulation based on the value of the logical expression. For example, an expression *Lathe.InputBuffer.Contents.NumberWaiting == 0* will have a selection value of 1 if the input buffer for the Lathe is empty, otherwise it will have a selection weight of 0.

Example: Routing by Link Weights

We will now build simple model of a TV inspection and adjustment process to illustrate the use of link weights for entity routing. Consider the following example in which TV's arrive to an adjustment process, and then undergo an inspection in which 20% fail inspection and are re-routed back for adjustment. We will have the rejected TV's turn red until re-inspection has been completed.

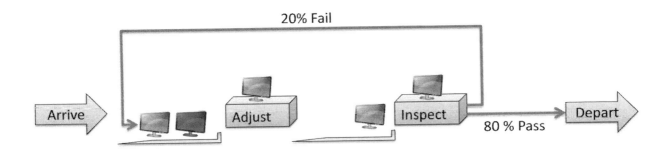

Open up Simio with a new project and place a Source, two Servers, and a Sink into the facility window. Rename the Source and Sink to *Arrive* and *Depart*, and the Servers to *Adjust* and *Inspect*, and draw paths connecting the output node of the Source to the input node of *Adjust*, the output node of *Adjust* to the input node of *Inspect*, the output node of *Inspect* to input node of *Adjust*, and the output node of *Inspect* to the Sink. Select the return path from the output node of *Inspect* to the input node of *Adjust* and set its selection weight property to 20, and select departure path from the output node of Inspect to the input node of the Sink and set its selection weight to 80. If you now click on the Run button you will see entities being randomly routed from the *Inspect* server based on the specified link weights.

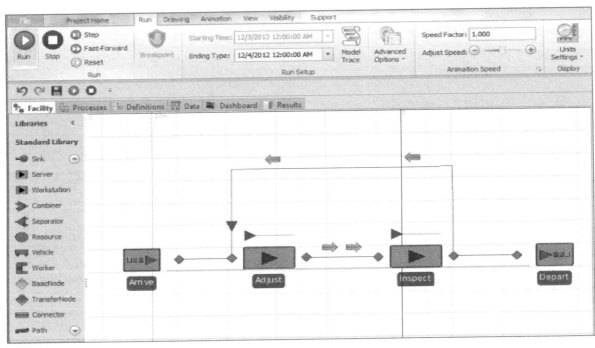

Next we will use Google warehouse to improve our animation by replacing the green triangles with a symbol of a TV, and also change the color of this symbol to red when a TV fails inspection, and then back to blue when a TV passes inspection. Drag out an instance of the *ModelEntity* from the Project library and place it in our facility. Select the entity instance, rename it *TV*, and with the TV selected click on **Download Symbol** in the **Symbols** ribbon and search for and select a *TV* symbol from Google Warehouse. With the *TV* symbol selected click on **Add Additional Symbol** in the Symbols ribbon; this adds a second symbol to the list of symbols for this entity. You can change between these symbols by selecting the drop list below Active Symbol and then clicking on the symbol you would like to be active. Make symbol 0 active and color it blue, and then make symbol 1 active and color it red.

To change between these symbols during the run we will make an assignment to the *ModelEntity* state variable named *Picture*. When *Picture* is set to 0 the entity will be drawn using symbol 0 (blue) and when *Picture* is set to 1 the entity will be drawn using symbol 1 (red). To change to the red symbol upon failure select the return path (*Path4*) from *Inspect* and click on the button in the *State Assignments -> On Entering* state assignment property to open the assignment editor. Click on Add button to add a new assignment, specify the state as *ModelEntity.Picture*, and specify the value as 1. This will cause the *ModelEntity* state *Picture* to be assigned a value of 1 upon entering the return path. To restore the entity to the blue symbol following adjustment click on the path departing Adjust, and click on the button in the *On Entering* state assignment property to open the assignment editor. Add an assignment to set *ModelEntity.Picture* to 0 (blue). This will cause all *ModelEntities* entering the departure path from Adjust to be displayed as blue. If you now rerun the model you will see the TV's changing between red and blue as desired.

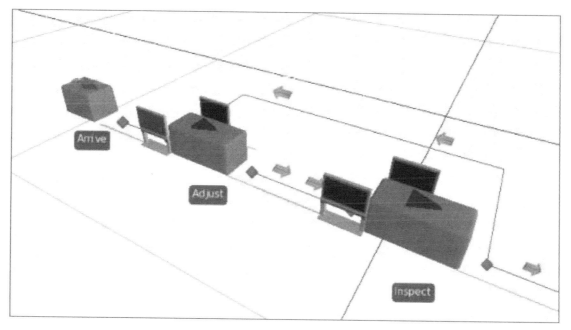

Setting the Entity Destination

Whenever the routing logic is based on the shortest path a destination is required for the entity. When using the Standard Library the destination for an entity may be set using the TransferNode by specifying the *Entity Destination Type* and its associated property values. Since the TransferNode is always used for the output side of the Standard Library fixed objects (Source, Server, Combiner, etc.), you can always specify the destination for an entity just as it departs any of the fixed objects in the Standard Library. To specify the destination you click on the TransferNode on the output side of the fixed object and specify the *Entity Destination Type*.

There are four options for specifying the *Entity Destination Type*. The first option is to *Continue* without reassigning the destination. This is the default option and is useful when the entity already has its destination set or when destination is not used.

A second option is *Specific* destination. In this case a required associated property is exposed for selecting the *Node Name* from the drop list. The node maybe be a simple name for a free standing node (e.g. *IntersectionA*), or a compound name (e.g. *Input@Server1* or *Output@Server1*) for a node that is attached as the input or output node of a fixed object. Compound node names are always specified using the format *NodeName@ObjectName*.

The third option for specifying a destination is *Select From List*. In this case we are dynamically selecting from a list of candidate destinations based on several criteria, which includes a *Selection Goal* (e.g. *Smallest Distance*, *Cyclic*, etc.), a *Selection Condition* that must be true for the node to be considered as a candidate for selection, and a *Blocked Routing Rule* that specifies if an "unblocked" node is required or preferred. An unblocked node is one that can immediately accept the entity. For example a Server with a finite buffer that is full would block any incoming entities.

The fourth option specifies that the destination is set *By Sequence*. In this case we have defined a specific visitation sequence (e.g. Drill, Lathe, Inspect, Depart) for each entity and we would like to set the destination to be the next node in this sequence. This is useful for modeling systems in which each entity has a different travel path through the system.

Regardless which method is used to set the destination for the entity, the shortest path routing logic can then be used to select the link that is on the shortest path to that destination. Note that although the use of the shortest path rule requires a destination, having a destination does not force you to use the shortest path rule: i.e. you can still employ link weights to control flow at one more locations along the way.

The *Select From List* and *By Sequence* options are both very useful and powerful methods for setting the destination for an entity. We will now illustrate each of these more advanced methods with an example.

Example: Select From List for Dynamic Routing

In this example we will use the *Select From List* routing option to send the arriving entities to the server with the fewest number of entities waiting in the queue. Consider the following system with arrivals being dynamically directed to one of three servers based on the smallest queue length. *Server1* has a constant processing time of 20 minutes. *Server2* has a random processing time in minutes with a triangular distribution and a minimum of 1, mode of 3, and maximum of 8. *Server3* has a processing time of .1 minutes.

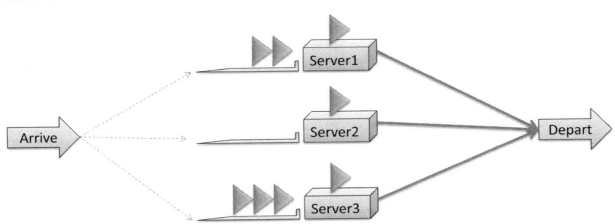

Open up Simio and place a Source, three Servers, and a Sink. Use a Path to connect the Source to the input node of each of the three servers, and also connect each of the Server output nodes to the Sink.

Edit the *Processing Time* for each of the servers (select the *Processing Time* in the Property window and click on the down arrow to open up the Expression Editor and begin typing) and also set the *Input Buffer* capacity for each Server to zero. This will force entities entering the Server to wait on the path since there is no input buffer available. Change the *Allow Passing* option on the Paths connecting the output node for the Source to the input nodes on the Server to *False* so that the entities will "queue up" along the path. Note that you can select the first, and then use Control-Click to add the second and third paths to the selection set, and then simultaneously change this property for all three paths. This same **Group Selection** approach can be used to simultaneously set the input buffer sizes to zero for the three servers.

Before using the *Select From List* option for selecting the destination we must first define our list of target nodes. In this case our target nodes are *Input@Server1, Input@Server2,* and *Input@Server3*. We will create a list of these nodes for selecting our destination. You could define a list on the Definitions window tab, and then click on the Lists icon in the panel selector on the left. But an easier way is to use the Group Selection technique described above to select the three gray target nodes, then right click and select **Add to Node List > Create** and name it Servers.

Now that we have defined our list, click on the blue output node for the Source to specify the routing logic from this node to the three Servers. In the Routing Logic section of the Property window change the *Entity Destination Type* to *Select From List*, and specify the *Node List Name* as the newly created node list named *Servers*. Make sure that the "Show Commonly Used Properties Only" option is not checked so you can use the full set of features. Set the *Selection Goal* to *Smallest Value* and keep the default Selection Expression as:

Candidate.Node.AssociatedStationOverload

Expressions are normally evaluated in the context of the parent object or the entity object executing the process logic, which in this case is our main model or the entity being routed. When we are searching the nodes in the list we want to tell Simio to use the context of the **candidate object** we are looking at in the search, and not the parent object or routing entity. The keyword *Candidate* is used for this purpose. The next part of the expression is telling Simio that the object type being examined is a class *Node*. The function *AssociatedStationOverload* is a built in function on the Node class that returns the overload at the associated station of this node. The overload is defined as the number of entities at this location, plus the number waiting to enter this location, plus the number of entities traveling with this node as their destination, minus the capacity at this location. Note that a positive number indicates the input location is currently overloaded (more there or on the way than the available capacity). A negative value indicates that capacity is still available based on those at or traveling to the Server.

Although this will generally balance the load between the three servers, we can make one additional change to further improve the balance by specifying the *Blocked Destination Rule* as *Prefer Available*. Note that we can have a situation where both *Server1* and *Server2* are busy, *Server3* is idle, but no one is currently travelling to any of the servers. By specifying the *Prefer Available* rule we will select *Server3* in this instance, which produces a more balanced flow of work to the three Servers. Our final model with these options set is shown below:

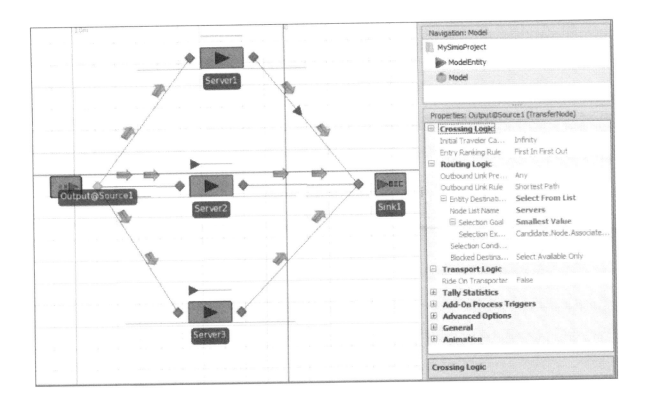

Example: By Sequence

In this example we will use the *By Sequence* option for setting the *Entity Destination Type* property. This option is useful for modeling situations in which an entity follows a specific sequence of destinations through the model. For example in a job shop each part type might have its own production sequence through the facility. To illustrate this functionality consider the following example where we have two different part types being processed across three different servers. Part type A is processed in the forward direction (*Server1, Server2, Server3*) and part type B is processed in the reverse direction (*Server3, Server2, Server1*).

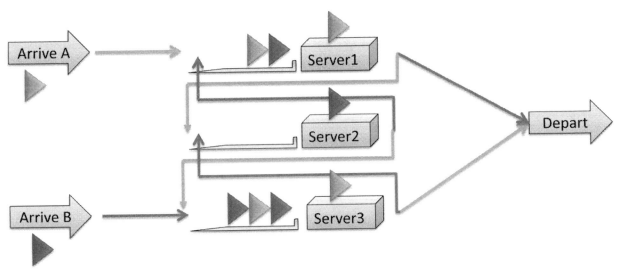

In this particular example the two part types do not share common paths. *PartA* will always travel on a distinct set of paths (shown in green), and *PartB* will always travel on a separate set of paths (shown in red). Hence in this case it would be possible to model this entity flow using two sub-networks. We would assign the green paths to *NetworkA*, and the red paths to *NetworkB*, and have the two part types follow their own sub-network through the system, creating the desired flow through the system. However this approach does not work in all cases: e.g., if *PartB* had a sequence *Server3-Server2-Server3*. Routing by sequence provides the flexibility needed to model these more complicated flows.

Begin by opening Simio, creating a new project, and placing the two Sources, three Servers, and Sink. Also drag out two entity symbols from the Project library and rename them *PartA* and *PartB*. Next we will color *PartB* red by clicking on the entity symbol, selecting the red color from the Symbols ribbon, and then clicking on the entity symbol to apply the color. Double click the path symbol in the Standard Library and draw the forward paths for *PartA*, and then continue drawing the backward paths for *PartB*. Change the *Entity Type* property for the second Source to *PartB*. At this point your model should appear as follows:

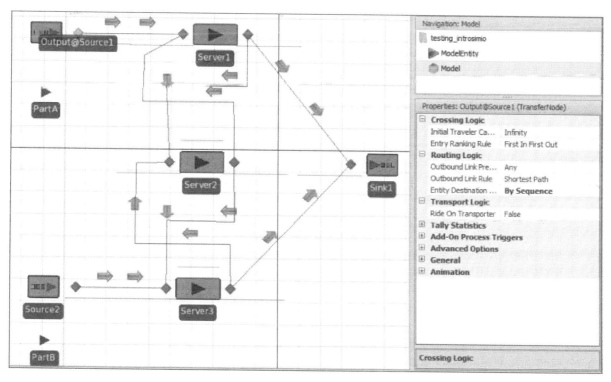

Our next step will be to create sequence tables for *PartA* and *PartB* that define their travel sequence through the system. To do this click on the Data window tab just under the ribbons. The Data window is used to define various types of data that can be used with your model and we will discuss this functionality in detail in Chapter 4. For now we will focus on creating Sequence Tables. In the tables section click on Add Sequence Table and rename the table *PartASeq*. Repeat this same step and rename this second table *PartBSeq*. Next drag the *PartBSeq* tab and drop it on the right window target so that we can see both tables simultaneously (you can alternatively right click on either table tab and select New Vertical Tab Group. Now enter the sequence *Input@Server1, Input@Server2, Input@Server3, Input@Sink1* for *PartASeq*, and enter *Input@Server3, Input@Server2, Input@Server1, Input@Sink* for *PartBSeq*. At this point your table data should appear as follows:

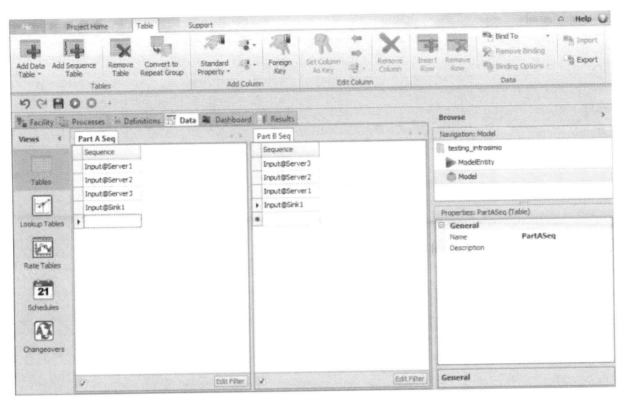

If you click on the column labeled Sequence in *PartASeq* the Property window will display properties for the sequence. The property *Accepts Any Node* is by default set to *True* and allows for any node in the model to be specified with its fully qualified name (e.g. *Input@Server1*). You can optionally change this property to *False* then only fixed objects may be specified using abbreviated names with the input node name and @ sign omitted for all fixed objects with a single input node (e.g. a Server). For example the sequence for *ParASeq* could be specified as *Server1, Server2, Server3, Sink1*. This is particularly convenient when importing routings from external data sources (discussed later).

Now that we have defined our routing sequences we are ready to use them in our model. Click on the Facility tab to bring the facility window to the front. Click on the *PartA* entity and set its Initial Sequence property to *PartASeq*, and click on the *PartB* entity and set its Initial Sequence property to *PartBSeq*. Now click on the output nodes for the two Sources and three Servers and specify *the Entity Destination Type* as *By Sequence* (you can do this by group selecting with Control-Click). Clicking on the Run button in the Run ribbon will now start the model executing.

Whenever an entity is routed from a node with the *By Sequence* option, it automatically bumps its index into the sequence. Hence each part type will be sent to the next destination that is listed in the sequence table that it is following. As we will see later you can also change the sequence table that an entity is following during the simulation (e.g. after failing inspection a part may be assigned to follow a rework sequence).

Summary

This chapter has focused on the concepts related to controlling the flow of entities between objects across a network of links. These basic concepts can be used to model a wide range of both static and dynamic entity flows. In the next chapter we will focus on the logic that executes once an entity enters the objects in the Standard Library.

Chapter 6: More about the Standard Library

Overview

In this chapter we will describe in more detail the objects in the Standard Library. This library is designed to be a generic modeling library with a broad spectrum of application. We have already made some use of the Source, Server, Sink, and Path objects from this library in the simple models that we have built so far. We will describe these in more detail, as well as the library objects that we have yet to use. However before describing these individual objects we will first introduce some general modeling concepts that are employed by these objects.

Preliminary Concepts

As noted earlier the objects in Simio have both a general behavior and specific behavior. The general behavior is defined by one of five basic classes: fixed, node, link, entity, and transporter. A **fixed object** models some general activities that take place at a specific location and are triggered by the arrival an entity. A **node object** models the starting and/or ending point of one or more links, and also defines the entry and exit point for a fixed object. Hence most fixed objects have associated node objects that provide input and output to the object. A **link object** models the pathway between two nodes. An **entity object** can be dynamically created and destroyed, may move through free space or over a network of links, and move into and out of fixed objects. Finally a **transporter object** is a sub-class of entity that can also pickup, carry, and drop off other entities. Although the object class defines the general behavior of each object, the specific behavior is defined by an internal model for the object definition. Each of these classes can have many different object definitions, each with different specific behaviors for the corresponding objects that are placed in a model. Hence the Standard Library is just one of many possible libraries that can be used for modeling. The ability to create custom libraries is a very powerful feature of Simio.

There are number of important concepts that are employed in designing and building object definitions. We will explore this topic in more detail in Chapter 6 where we explore the design and building of custom objects. For now it is important to understand some basic concepts related to **stations** and **resources**.

Many of the objects in the Standard Library make use of the concept of a station. A **station** is a construct in Simio that is used to model a constrained location within an object. Entities can transfer from station to station within an object. Only one entity can transfer into a station at a time, and this transfer can require time to execute. A station can also have a fixed capacity that will limit the number of entities that are in process at the station. Whenever a station is at its capacity the station will block any attempts by other entities to transfer into that station.

As an illustration, the Server object employs three stations named *InputBuffer*, *Processing*, and *OutputBuffer*. An entity first enters the Server at the *InputBuffer* station, then moves to the *Processing* station, and then finally moves to the *OutputBuffer* station. Each station in Simio has two internal queues named the *EntryQueue* and *Contents*. The queue named *EntryQueue* holds entities that are waiting to transfer into the station, and the *Contents* queue holds the entities that are currently located inside the station. Hence the expression *Server1.Processing.Contents* denotes the queue of entities in the *Processing* station at Server1, and the expression *Server1.Processing.EntryQueue* denotes the queue of entities waiting to transfer into the *Processing* station. Likewise the Vehicle object employs a station named *RideStation* to hold the entities that are currently riding on the Vehicle.

Many of the objects in the Standard Library also declare themselves as a **resource**. If an object declares itself as a resource, then it has a changeable capacity that can be seized and released by other objects. The object also as a queue named *AllocationQueue* that holds a prioritized list of other objects that are waiting to seize it. When capacity becomes available it can be allocated statically based on this prioritized list, or dynamically based on user-customizable rules. This latter feature makes for very flexible resource allocation. As we will see later, objects can implement logic to either accept or reject a seize attempt by another object. When an object is seized a visit request can also be made to have the seized object travel to a specific location. In this case the request is placed in the *VisitRequestQueue* for the object and the object can then decide when and if to visit the requesting object. Hence seizing and moving a resource object requires cooperation between the seizing object (e.g. a server) and the object being seized/moved (e.g. an operator).

Within the Standard Library most of the fixed objects, nodes, and the Vehicle and Worker are declared as resources. You can also add entities and transporters to your project and declare them as resources. The Standard Library also contains a fixed object named Resource that is specifically provided for use as a stationary (non-moving) resource object for seizing and releasing.

When using the Design or Team Edition of Simio the objects in the Standard Library support the concept of Add-On processes. An Add-On process is user-defined process that may be inserted into the object logic at a number of predefined points. As a result you can actually customize the behavior of these objects without having to create new objects. We will discuss this functionality in detail in Chapter 9. Most of the Standard Library objects also support the concept of entity state assignments upon arrival to and/or departure from the object.

Source and Sink

As we have already seen in our previous examples, the Source and Sink are fixed objects that are used to dynamically create and destroy entities. We will now exam each of these objects in more detail.

The **Source** is comprised of the main Source object plus the associated output node (type TransferNode) named *Output*. You independently select and edit both the main Source object and the associated output node. The main Source object has an *OutputBuffer* station that holds entities that are waiting to exit the source through the associated output node. Once an entity successfully transfers into the network through the output node it then departs the *OutputBuffer* station that is owned by the Source. The relationship between the OutputBuffer station that is owned by the Source, and the associated Output node is shown below:

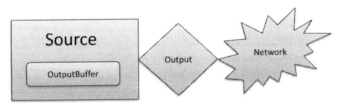

By default the Source includes an attached animated queue that displays the entities in its *OutputBuffer.Contents* queue. Note that an animated queue can be added as either a detached queue or an attached Queue. An attached queue is one that is defined in the context of and moves with the associated object, whereas a detached queue is free-floating and defined in the context of the parent model. Hence if you drag the Source object the attached animated *OutputBuffer.Contents* queue will move with it. If you place a Source, click on the output queue, and press the delete key you will delete the animated queue. Note that this does not in any way impact the logic for the Source – the logical station queue still exists – its contents are simply no longer being animated. To re-attach an animation queue to the Source click on the Source object in the facility, click on the down arrow on the Queue icon, and select *OutputBuffer.Contents* from the drop list. To add a detached queue select the Animation ribbon and click on the Detached Queue icon. In this case you must enter the Queue State property to animate using the fully qualified name (e.g. *Server1.OutputBuffer.Contents*).

The arrival logic properties for the main Source object include the *Entity Type, Arrival Mode,* and *Entities Per Arrival*, along with additional properties that switch in and out based on the *Arrival Mode*. The *Entity Type* specifies the type of Entities created by the Source. The *Entities Per Arrival* property specifics how many entities are to be created for each arrival event at the source.

The *Arrival Mode* property governs the pattern of arrivals. The default mode is specified as *Interarrival Time*, which switches in properties for specifying the *Time Offset* for the first arrival, and *Interarrival Time* for subsequent arrivals. The mode *Time Varying Arrival Rate* is used to model a non-homogenous Poisson arrival process in which the time varies according to a Rate Table that is defined in the Data window. This is often used to model customer service where demand changes over the course of the day (e.g., busy lunch hour and mid-afternoon lull). To create a Rate Table select the Data window, click on Rate Tables in the selection panel on the left, and then click on Rate Table in the ribbon, which then creates a default Rate Table. A Rate Table is a set of equal size intervals that specify the mean arrival rate during that interval. Note that the arrival process is random and so that actual number of arrivals during a given interval may be more or less than this expected value. You can specify the number of intervals as well as the interval size. The Rate Table will automatically repeat once it reaches the end of the last interval. The following shows a Rate Table with six intervals, each four hours long. Once this Rate Table is defined, it can then be referenced with the *Time Varying Arrival Rate* mode.

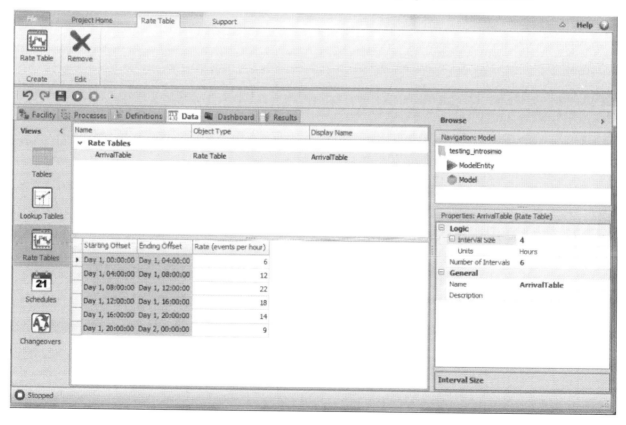

The mode *Arrival Table* is used to generate a specific set of arrivals based on arrival times specified in a table that is defined in the Data window. To use this mode we must first create a table that contains a list of arrival times that entities enter the system. To create an Arrival Table select the Data window, click on Tables in the selection panel on the left, and then click on Add Data Table in the ribbon, which then creates a default Data Table. We will rename this table by changing the *Name* property in the Property window to *CustomerTable*. We will now click on the down arrow for Standard Property and select a Date Time property to add to the table, and rename this property ArrivalTime. We can enter arrival time values into this table as shown below:

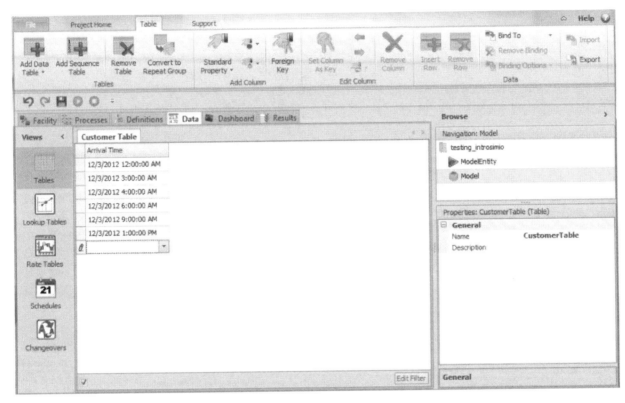

Once we have defined our arrival table we can then reference this table and arrival time column using the Arrival Table mode for the Source object. In this example the Arrival Table is specified as *CustomerTable.ArrivalTime*. The Source will then generate arrivals at the times specified in this table. You can also apply a random deviation to the scheduled arrival times, as well as a probability of "no show". We will discuss the creation and use of Data Tables in much greater detail in Chapter 7 where we describe the use of tables for building data driven models.

The *On Event* mode is used to create a new entity every time an event fires, and switches in a property for specifying the *Event Name* for the triggering event. Many of the objects in the Standard Library fire events at selected times, and we can use these to cause the creation of a new entity. For example specifying the triggering event as *Input@Sink1.Entered* would cause the Source to create a new entity each time an entity entered the input node at Sink1. When selecting this mode you can also specify the property *Initial Number Entities* to create an initial set of entities in addition to those that are created for arrivals that are triggered by each event firing.

Source Stopping Conditions – Note that the Source object also provides a 'Stopping Conditions' category of properties that allows you to easily stop arrivals generated by the source using one or more of the following conditions:

- After a specified Maximum Arrivals has occurred.
- After a specified Maximum Time has elapsed.
- After a specified event has occurred.

The **Sink** object is the counterpart to the Source object – it destroys arriving entities and removes them from the system. The Sink is comprised of the main Sink object plus the associated input node (type BasicNode) named *Input*. Only one entity at a time can be processed by the Sink. The Sink has an InputBuffer station that holds the entity being processed for the specified *TransferIn Time*. Entities are transferred from the network thru the *Input* node and into the *InputBuffer* station. The relationship between the *Input* node and *InputBuffer* station is shown below:

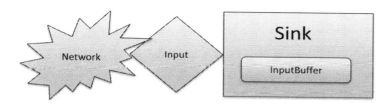

The Sink comes with an attached queue for animating the entity in the *InputBuffer* station. The Sink automatically records the average time in system for entities that are destroyed by the Sink.

Server

We have used the Server object in several of our previous examples to model a constrained service operation. The main Server object has two attached objects: an input node (type BasicNode) and an output node (type TransferNode).

As noted earlier the main Server is modeled with three stations: *InputBuffer, Processing*, and *OutputBuffer*. An arriving entity typically enters from the network through the *Input* node to the Server's *InputBuffer*, proceeds through *Processing*, and then waits in the *OutputBuffer* to move through the output node and into the network. Note that the network may be blocking an attempt to transfer out through the *Output* node (e.g. a path with no passing may be fully backed up and have no available space). In a similar way when the Server is full it may be blocking the network on the input side.

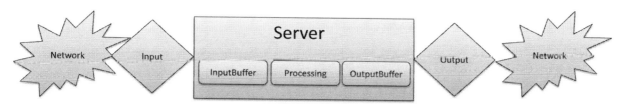

Both the *InputBuffer* and the *OutputBuffer* stations can have their capacities set to zero which causes them to be removed from the logic. Hence if the *OutputBuffer* capacity is specified as zero the entity completing *Processing* will wait in *Processing* to move through the output node and into the network. Likewise if the *InputBuffer* capacity is set to zero, then entities that are entering from the network through the input node must proceed directly to the Processing station of the Server. If the Processing station is at its capacity it will block entities attempting to enter the Server.

Whenever processing capacity becomes available the next entity to be processed is selected from the *InputBuffer*. The *InputBuffer* is ranked based on the Ranking Rule property, that can be specified as *First In First Out, Last In First Out, Smallest Value First, or Largest Value First*. In the last two cases you specify an expression on which the ranking is based – e.g. you could rank the queue based on processing time or due date. If no *Dynamic Selection Rule* is specified then the first entity in the statically ranked queue is selected for processing. However if a *Dynamic Selection Rule* is specified, then all entities in the queue are examined and considered for selection based on the specified rule. These rules include *Smallest Value First* and *Largest Value First*, which from their name seem like similar options to the simple *Ranking Rule*. However this dynamic rule is actually more flexible and powerful (although slower to execute) because it re-evaluates the expression for each entity at the time each decision is made. For example an expression that has a term that computes if the entity is tardy (by comparing its due date to the current time) would correctly use the current time when making this decision, and not the arrival time to the queue. The three campaign rules are used to make selections in an increasing or decreasing order based on a specified expression. For example, a Server modeling a painting operation may move from light to dark colors, and then start over with the lightest color once no darker colors remain.

The *Transfer-In Time* property on the Server specifies the time required for an entity to transfer into the Server object. Only one entity may transfer at a time, and processing of the entity may not start until this transfer is completed. The *Processing Time* property specifies the time required to process the entity at the Processing station.

The Server object can be used to model either a single server or a single processing center with multiple identical servers, depending upon the capacity specified for the Processing station. The Processing station is governed by a *Capacity Type* that is specified as either a *Fixed* capacity, or based on a *WorkSchedule*. Note that a single work schedule can be referenced by multiple servers. To specify a capacity based on a *WorkSchedule* first create a schedule by clicking on the Data window, selecting the Schedules panel on the left, and then clicking on Create Work Schedule in the Schedule ribbon. To define a new schedule you must first create one or more day patterns that define a pattern of work for a specific day. For example we might define a *Standard Day* as a day pattern comprised of work from 8am to noon and 1pm to 5pm, or a *Half Day* as a day pattern comprised of work from 8am until noon. Each work period in a day pattern is defined by a start and end time, a capacity value, and an optional description. By default the schedule is off shift, so within the day pattern you only need to define the capacity for on shift periods. Day patterns are then used as building blocks to create work schedules. The following shows a simple day pattern for a Standard Day with two work periods from 8 to 12 and 1 to 5.

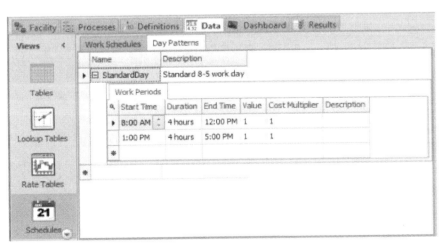

A work schedule is made up of two parts: a repeating work cycle defined by a repeating set of day patterns with superimposed work exceptions. A work cycle defaults to seven days, but can be set to any arbitrary number of days. You specify a day pattern (e.g. *Standard Day*) for each work day in the repeating cycle.

A work exception is used to override the repeating work pattern. Work exceptions can be added to a work schedule by clicking on the expand button next to the work schedule name to expose the exception tables. You can add Work Day Exceptions to replace an entire day pattern on a specific date (e.g. a holiday), or Work Period Exceptions to specify an exception that covers a portion of a single day or spans one or more days (e.g. from 2pm on Tuesday to 4pm on Friday). The following shows a seven day work schedule named *Standard Week* with *Standard Day* work patterns for Monday thru Friday, and a work day exception (no day pattern) for the fourth of July.

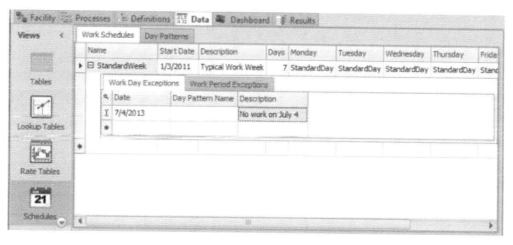

When modeling multiple servers there is always the option to model each server as a separate Server object in Simio, or use a single Server object with a capacity set to the number of servers. The latter is far simpler and will produce identical results as long as all the servers have the same performance characteristics, and waiting customers always go to an available server.

The Server also provides options for failures that halt processing by the Server during the failure. The failures supported include *Calendar Time Based, Processing Count Based, Processing Time Based,* and *Event Count Based*. In all cases the failure duration is specified by the *Time To Repair* property.

Entities passing through the Server will always seize the Server as its primary resource that is required for processing the entity. It is also possible to specify a secondary resource (tooling, operator, etc.) that is required during processing. This resource, if specified, is seized before processing starts and is then released once processing completes. If the specified resource is a moveable resource (e.g. a Worker object) the resource may optionally be required to visit a node associated with the Server before processing may begin (we will see an example of this later in this chapter). Additional flexibility is provided by allowing other resources to be individually seized and/or released at selected points within Server. Optional seize and release points are provided on entering the Server, before processing on the Server, and after processing on the Server. Note that this also allows resources to be seized at one Server, and then released at a second Server.

By default the Server object has animated queues for the *InputBuffer, Processing,* and *OutputBuffer*. Again these are animation only constructs and have no impact on the logic of the Server. It may be necessary to lengthen the animation queue to see all the entities that are actually in the station.

Combiner and Separator

The Combiner and Separator are similar to the Server in many ways, however provide additional functionality to combine several entities together, and then to separate them or to make copies of a single entity.

The **Combiner** is used to model a process that matches multiple entities, groups those entities into a batch, and then attaches those batched members to a parent entity. For example, if we have parts being placed on a palette, the batched members would be the parts, and the parent entity would be the palette. The individual batched entities can later be split out using the Separate.

The Combiner is comprised of a main object plus three associated node objects, including an input node named *ParentInput* for the parent entities, an input node named *MemberInput* for the member entities, and an output node named *Output* for the batched entities. The main Combiner object has four stations: *ParentInputBuffer, MemberInputBuffer, Processing,* and *OutputBuffer*. Arriving parent entities transfer from the network thru the *ParentInput* node to the *ParentInputBuffer* station, and arriving member entities transfer from the network thru the *MemberInput* node to the *MemberInputBuffer* station. The entities are then formed into batches in the *Processing* station, and then wait in the *OutputBuffer* station to depart into the network through the *Output* node. The relationship between these node objects and the associated stations is shown below:

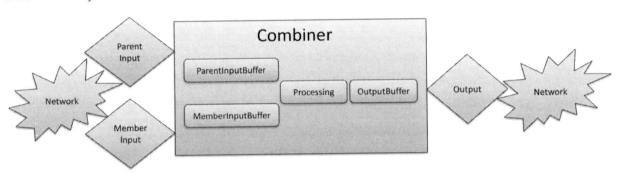

The properties for the Combiner include the *Batch Quantity* and *Matching Rule*. A batch is formed by selecting the number of members specified by the *Batch Quantity* from the member queue, and attaching them as a batch to a single parent entity selected from the parent queue. The *Matching Rule* options include *Any Entity, Match Members,* and *Match Members and Parent*. In the case of *Any Entity* then entities are selected in the order that they are ranked in the queue. In the case of *Match Members* then a batch can only be formed if the member entities all have the same value for the *Member Match Expression*. In the case of *Match Members and Parent* then a batch can only be formed if the member entities all have the same value for the *Member Match Expression*, and there is a waiting parent entity with the same value for the *Parent Match Expression*. The batched members are carried by the parent entity in a queue named *BatchMembers* which can be animated on the parent entity symbol.

The processing capacity, reliability, and secondary resources for the Combiner are similar to the Server. A good way to think about the Combiner is that it's a special purpose Server with two input nodes and special processing logic for batching.

The **Separator** is used to model a process that separates batched members from a parent entity, or makes copies of an entity. Essentially the Separator is a mirror image of the Combiner. The Separator is comprised of a main object plus three associated node objects, including an input node named *Input* for the arriving parent entities, an output node named *ParentOutput* for the departing parent entities, and an output node named *MemberOutput* for the departing member entities. The main Separator object has four stations: *InputBuffer*, *Processing*, *ParentOutputBuffer,* and *MemberOutputBuffer*. Arriving parent entities transfer from the network thru the *Input* node to the *InputBuffer* station. The entities are then split apart or copied in the *Processing* station, and then the parent entities wait in the *ParentOutputBuffer* station to depart into the network through the *ParentOutput* node, and the member entities wait in the *MemberOutputBuffer* station to depart to the network thru the *MemberOutput* node. The relationship between these node objects and the associated stations is shown below:

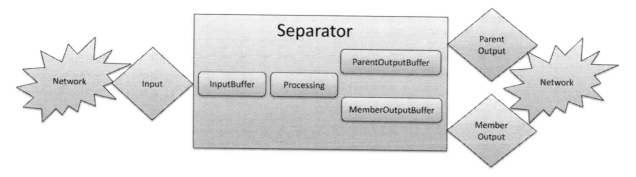

The processing logic for the Separator is controlled by the *Separation Mode* that can be specified as either *Split Batch* or *Make Copies*. In the case of the *Split Batch* option, entities that are carried by the parent entity in its BatchMembers queue are split off and transferred to the MemberOutputBuffer where they wait to depart the Separator. The parent entity is then transferred to the ParentOutputBuffer when it waits to depart the Separator. The number of entities that are split off is specified by the Split Quantity property which defaults to the total number of members in the parent entity's BatchMembers queue. In the case of the *Make Copies* option members are created by cloning the parent entity, with the number of newly created members specified by the *Copy Quantity*.

Example: Combine then Separate

In this example we will model the arrival of two different entity types that have independent arrival processes and are combined together, processed as a combined batch, and then un-batched before departing the systems. Each entity type arrives randomly with an exponential time between arrivals with a mean of 1 minute. The time to batch two entities is a constant .5 minutes, and the time required to un-batch the entities is a constant 1 minute. The processing time for the combined batch has a triangular distribution with a minimum of .5, mode of .8, and maximum of 1.2 minutes.

Begin by opening Simio and selecting a new model Project. Now place two entities from the Project Library and rename them *Parent* and *Member*. Resize the *Parent* entity to make it larger. Select the *Parent* entity and click on the down arrow below the Queue icon on the Symbols ribbon and select *BatchMembers*. Using the cross hair cursor draw the attached queue just above the *Parent* entity (make sure it's long enough to contain at least one *Member* entity) and then terminate queue drawing with a right-click. Note that this attached queue will animate any member entities that are held in the BatchMembers of the Parent entity. Now select the *Member* entity, select the color red, and click on the *Member* entity again to apply the color. Drag out two Sources, a Combiner, a Server, a Separator, and two Sinks from the Standard Library. Now connect the objects with Paths as shown below:

Introduction to Simio

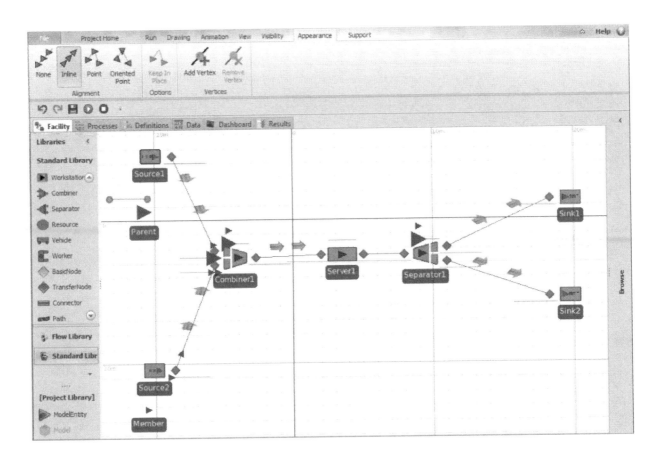

Change the *Interarrival Time* for both Sources to *Random.Exponential(1)*, and change the *Entity Type* for Source2 to *Member*. Set the *Processing Time* properties for the Combiner, Server, and Separator to *.5*, *Random.Triangular(.5, .8, 1.2)*, and *1* minute respectively. Now run the model to watch the batching and un-batching process.

Basic Node and Transfer Node

The Standard Library contains two different object definitions of type node. The Properties for these nodes were described in detail in Chapter 2. The **BasicNode** models a simple capacitated intersection with both *Shortest Path* and *By Link Weight* routing logic. The **TransferNode** provides additional functionality to support setting/changing the entity destination as well incorporating transport logic to support riding on transporters. We will discuss this option in more detail later in this chapter. Both the BasicNode and TransferNode also allow you to record statistics on entering and/or exiting the node.

Within the Standard Library the BasicNode is used as an associated object on the input side of a fixed object, and the TransferNode is used as an associated object on the output side of a fixed object. Although these node objects are mostly used as associated objects with other fixed objects, these can be also be free-standing objects that are used to model intersections and even in-line processing locations in a network of links and nodes.

Nodes have a parking station named *ParkingStation* where entities/transporters can pull into and out off at the node location. A parking station allows entities to be associated with a node in the network, but be removed from the crossing area of the node so that other entities can pass through without constraint.

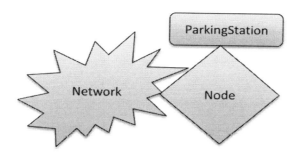

The entities in the parking station at a node can be animated by clicking on the node, clicking on the down arrow on the Queue icon in the Appearance ribbon, and adding an animated queue to display the entities in ParkingStation.Contents.

Connector, TimePath, and Path

Three commonly used links in Simio include the Connector, TimePath, and Path. The **Connector** is used to transfer an entity between two nodes in zero time. The only property on the Connector is the *Selection Weight* used for routing *By Link Weight*. The **TimePath** is used to transfer an entity between two nodes with a specified *Travel Time*. The TimePath can also have a limited *Travel Capacity*, and an *Entry Ranking Rule* for controlling the order in which entities enter the TimePath.

The **Path** is used to model a pathway in which each entity on the Path can move at its own assigned speed. As a result the Path has an *Allow Passing* property that governs what happens when a fast moving entity closes in on a slower moving entity. The Path also has a *Speed Limit* property that will limit the speed of any entities moving thru the Path. By default the travel of the path is based on the geometric length of the pathway that is drawn in the facility. If you graphically increase the length of the physical path then you also increase the logical travel distance. You can also set the *Drawn To Scale* property to false which allows you to specify the *Logical Length* of the Path.

Both the TimePath and the Path have a *Type* property that can be specified as *Unidirectional* or *Bidirectional*. If you set this property to Bidirectional then entities can travel through the link in either direction. However with *Bidirectional* links traffic can only travel in one direction at a time. For example if a *Bidirectional* link is drawn between nodes A and B, then any traffic traveling from B to A must wait for all traffic traveling in the link from A to B to complete their travel and exit the link.

With *Bidirectional* links it is very easy to create situations where the traffic will deadlock. The link network should be designed in such a way that deadlocks are avoided.

Example: Bidirectional Paths

In this example we have two different parts that travel in opposite directions through a bidirectional network with sides provided to avoid deadlock. Our example model is shown below. *PartA* entities enter at *Source1* and have their destination set on the Output node for *Source1* to *Sink2*. *PartB* entities enter at *Source2* and have their destination set on the Output node for *Source2* to *Sink1*. We have placed four free-standing BasicNodes vertically aligned in the center of the facility to represent intermediate intersections. We have drawn Paths between nodes as shown and set *Allow Passing* to *False* on all Paths. We have also set the two vertical Paths in the center of the facility to *Bidirectional*.

Note that entities will automatically wait for oncoming traffic to clear before entering a *Bidirectional* link. Note also the crucial role played by the two side pathways in the center of the facility to buffer against deadlocks with the two *Bidirectional* paths. Without these side pathways traffic would deadlock at the intersection of the two *Bidirectional* paths.

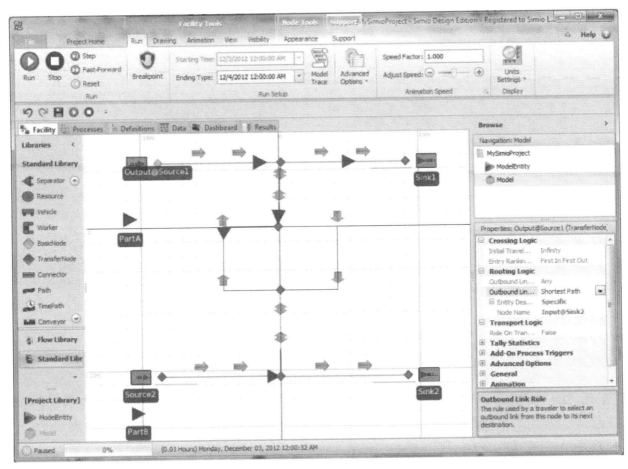

Conveyors

The fourth type of link provided in the Standard Library is the **Conveyor**. In contrast to the Path where the entities can move independently at different speeds across the link, with a Conveyor the entity movement is governed by the current speed of the Conveyor. Entities merge onto and consume space on the conveyor and rely on the conveyor to provide their movement. The *Accumulating* property determines what happens with the Conveyor when an entity reaches the end of the link and is blocked from continuing. If *Accumulating* is *True* then the conveyor does not stop moving and the moving Conveyor is assumed to slip relative to the stopped entity. In this case trailing entities will accumulate behind the stopped entity. If *Accumulating* is *False* then the Conveyor is also stopped. The Conveyor will start moving again at the *Desired Speed* once the blockage is removed.

By default entities can merge onto the Conveyor at any location. Some conveyor devices (e.g. bucket conveyors, ski lifts, etc.) only allow entities at fixed locations along the device. This can be modeled by changing the *Entity Alignment* property from *Any Location* to *Cell Location*, and then specifying the *Number of Cells* or *Cell Size* on the Conveyor.

Example: Merging Conveyors

In this example we have two different types of boxes arriving along separate accumulating conveyors to a merge point, where they continue along a third accumulating conveyor to a server. Following a labeling operation they continue on a fourth accumulating conveyor to their departure point from the system. Both box types arrive with an interarrival time is that is exponentially distributed with a mean of .5 minutes. The service time has a triangular distribution with a minimum of .1, mode of .2, and maximum of .3 minutes. The conveyor supplying Box1 moves at 3 meters per second, and the conveyor supplying Box2 moves at 4 meters per second, and these merge onto a conveyor moving at 2 meters per second. The boxes depart from the server on a conveyor moving at 2 meters per second. The facility layout is shown below:

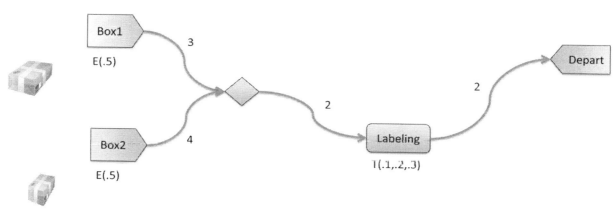

Begin by opening Simio and selecting a new model Project. Place two Sources, a Basic Node, Server, and Sink and connect them with conveyors. Now place two entities from the Project Library and rename them *Box1* and *Box2*, and select new symbols for the entities from the Project Symbols on the Symbols ribbon. Edit the parameters for the *Interarrival Time* on *Source1* and *Source2*, and change the *Entity Type* for *Source2* to *Box2*. Set the *Input Buffer* capacity to 0 for *Server1* (which causes entities to accumulate on the conveyor at the input to the Server), and change the conveyor speeds coming out of the Sources to 3 and 4, respectively. Click on each conveyor and apply the Path Decorator for the conveyor. The model is now ready to run and shown below:

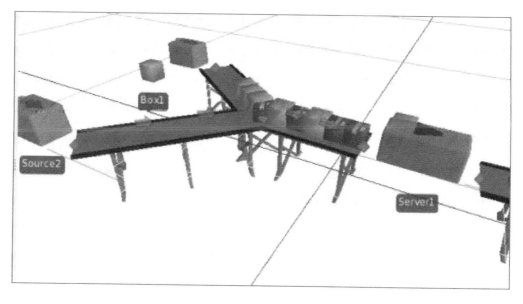

Vehicle

The Standard Library has a transporter definition named Vehicle. The **Vehicle** may be used to transport objects between node locations in the network. A transporter has a single station named RideStation where entities are held during transport. The Vehicle can be used to model devices that follow a fixed route (e.g. a bus, train, etc.), or respond to dynamic requests for pickups (e.g. taxi, AGV, etc.). To use the Vehicle you drag a Vehicle object from the library and place it anywhere in the facility model. You then click on this Vehicle object to edit its properties.

The routing logic for a Vehicle is controlled by properties that specify the *Initial Node* and *Routing Type*. The *Initial Node* (also referred to as the Home node) specifies the starting location for the transporter. The *Routing Type* is used to specify if the transporter follows a *Fixed Route*, or responds to *On Demand* request for pickups and drop-offs. In the case of *Fixed Route* the sequence table name defining the *Route Sequence* is also specified. In the case of *On Demand* the *Idle Action* to take when there are no active pickups or drop-offs to perform is also specified. The options for *Idle Action* and *Off Shift Action* include *Go To Home* (travel to the Home node but remain on the network), *Remain In Place* (i.e. remain at current location on the network), *Park At Home* (travel to the Home node and pull into the parking area at that node), *Park At Node* (pull off the network into the parking area for the current node), and *Roam* (move randomly through the network based on link weights).

Note that when modeling with transporters it is typically desirable to add animated queues for the parking stations at each node where the transporter may park. This would include the *Initial Node* where the transporter starts, as well as any nodes where the transporter may park along the way. If you do not add these animated queues to these nodes then the transporters will show up in default parking queues whenever they are parked in the parking station.

By default, a parking area is supplied adjacent to each node. Whenever an object (such as a vehicle) parks at the node, if the Parking Queue option is enabled for the node on the Appearance ribbon, then the object will be displayed in the default parking queue. If you want to customize parking animation in any way, then you may disable this option and use the Draw Queue button in the Appearance ribbon to manually add the node's ParkingStation.Contents queue animation.

The transport logic for a Vehicle is defined by the *Initial Ride Capacity, Load/Unload Times*, and *Task Selection Strategy*. The *Initial Ride Capacity* specifies the maximum number of passenger entities that can be carried by the transporter. The *Load/Unload Times* specify the time required to either load or unload each passenger that gets onto/ off of the transporter. The *Task Selection Strategy* specifies a rule that the transporter uses to select its next pickup or drop-off task to perform. The strategies supported include *First In Queue, Smallest/Largest Distance,* and *Smallest/Largest Priority*. The *Minimum Dwell Time Type* specifies the minimum amount of time (if any) a vehicle is required to 'wait' at a node to load or unload entities. Options include *No Requirement, Dwell Until Event, Dwell Until Full,* and *Specific Time*.

The *Initial Number In System* property in the *Population* category on the Vehicle can be used to specify the initial number of vehicles in the system. This quantity of vehicles will be created and placed into the parking station for the *Initial Node* at the beginning of each run. Note that it is also possible to dynamically create and destroy vehicles as the model is running.

When using a transporter you must specify that entities at specific pickup points are to ride on the transporter. You can do that using the TransferNode in the Standard Library. Note that the TransferNode is used for the output node of the fixed objects (e.g. Source or Server) and this makes transporters available for access at these locations. To specify that the departing entity is to ride on a transporter (as opposed to moving through the network on its own) you set the *Ride On Transporter* property for this node to *True*, and then either specify a *Specific* transporter to use, or select the transporter *From List*. In the latter case you also specify a *Reservation Method* for reserving a pickup when there is no transporter at your current location, and a *Selection Goal* and *Selection Condition* for selecting between available transporters that are currently at your location. Note that the *Reservation Method* is only employed if there is no local transporter to select. The available *Reservation Methods* include *Reserve Closest, Reserve Best, and First Available at Location*. The *Reserve Best* option applies the selection goal and condition that is used for selecting local transporters. Note that transporters will be travelling through the network performing drop-offs for its riders, and also pickups based on reservations that have been made. The *First Available at Location* waits for a transporter with available capacity to pass through this location and makes an immediate reservation with that transporter.

By default a transporter with available space will always pick up a rider, and an entity will also ride on an available transporter. As we will see in Chapter 8 when using the Design and Team Edition of Simio you can employ an add-on process for a transporter to evaluate transport requests and refuse giving rides to specific entities. Likewise you can employ an add-on process for an entity to evaluate and reject a candidate transporter. This gives added flexibility for controlling how transporters are used within the model.

A Vehicle object with *Routing Type* specified as 'On Demand' can also seized/released by other objects for processing tasks. Each vehicle has a resource capacity of 1 and thus may be seized and used for one task at a time. When seized, a visit request to a specified node location can also be made for a Vehicle. Note that, if being used for both processing tasks as well as transport tasks, a vehicle will always prioritize seize visit requests over any new transport pickup requests. Hence the Vehicle has dual modes: as a moveable resource (e.g. similar to a worker) that is seized and released and travels from location to location in the model, and as a transporter that can pickup, carry, and drop off entities at different locations.

Example: On-Demand Pickups

In this example passengers arrive to the system with an inter-arrival time that is exponentially distributed with a mean of 5 minutes. The passengers are picked up at the arrival point by a vehicle that can hold up to two passengers at a time, and transports them to the entry location for a service area, where their service time in minutes has a triangular distribution with a minimum of 1, mode of 2, and maximum of 3. Once passengers complete service they are then picked up by the same vehicle at the service area's exit location and carried to the departure point. The vehicle requires .3 minutes to load or unload a passenger and can travel 2 meters/second on pathway that has both unidirectional and bidirectional links as shown below.

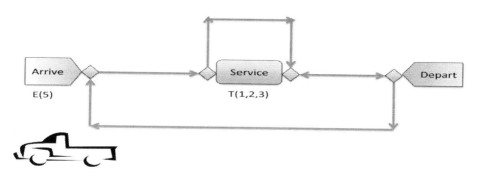

Begin by opening Simio and selecting a new model Project. Now drag out a Source, Server, and Sink and connect them with Paths as shown, changing the *Type* for Path from the Server to the Sink to *Bidirectional*. Edit the *Interarrival Time* for the Source and *Processing Time* for the Server, and set the *InputBuffer* capacity for the Server to 0 (causing the Server to block the Vehicle on drop-offs).

Now drag an instance of the Vehicle into your model. Set the *Initial Node* for *Vehicle1* to *Output@Source1* and change the *Idle Action* to *Remain In Place* (which causes the Vehicle to sit at its current location when it goes idle). Also change the *Ride Capacity* to 2 and set the *Load* and *Unload Time* to .3. Finally change the *Ride On Transporter* property to true and the *Transporter Name* to *Vehicle1* on the output nodes on the Source and the Server. The model is now ready to run and is shown below:

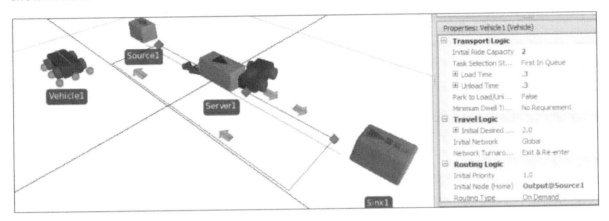

Workstation

The **Workstation** is similar to a Server with capacity one, except that it models the *Processing* station in far more detail. The *Processing* station is represented by an operation that is divided into three distinct activities: *SetupActivity, ProcessingActivity,* and *TeardownActivity*. Each entity moving through the Workstation will perform each of these activities within the Processing station.

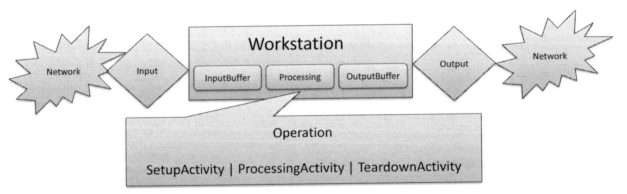

The entities passing through the Workstation represent a production lot, where the number of items in the lot is specified by the *Operation Quantity*. There is a single setup activity required before starting production for the lot, and a single teardown activity performed once the production lot is completed. During processing the production lot may be divided into smaller production batches that are sequentially processed.

The time required for the *SetupActivity* is determined based on the *Setup Time Type*, which can be specified as *Specific, Change Dependent,* or *Sequence Dependent*. The last two options dynamically calculate the setup time based on an *Operation Attribute*. The *Operation Attribute* is an expression evaluated for the entity that typically involves one or characteristics (e.g. color, size, etc.) that are selected from a list. In the case of a *Change Dependent* setup we have two possible setup times – one if the *Operation Attribute* expression remains the same (e.g. same color as last operation), and one if this value changes. In the case of a *Sequence Dependent* setup the time is given by a Changeovers matrix that is defined in the Data window. To create a Changeover matrix you must first define a list of possible values (e.g. *Small, Medium*, and *Large*, or *Red, Green, Blue*), and then specify that list for use in defining the matrix. In this case each possible changeover (e.g. from *Small* to *Large*) has a different setup time.

Once the *SetupActivity* is complete the *ProcessingActivity* is started. By default the entire production lot is processed as a single batch, however you can also specify a smaller batch size using the *Processing Batch Size* property.

The *Other Requirements* category includes properties for specifying additional constraints on the production process. Clicking on the repeating dialog button in the *Secondary Resources* property opens up the Secondary Resources – Repeating Property window where you can define one more additional resources that are required during one or more activities. For example you might require an operator during setup and teardown, and a tool during the entire operation. You can specify a specific resource that you require, or you can select a resource from a list using a selection goal and condition. In the case of a moveable resource (e.g. a Vehicle or Worker) you can request that the resource visit a specific node before the activity can begin. For simple fixed resources you can place instances of the Resource definition into your model and employ these as secondary resources.

Each batch that is processed can both consume materials at the start of processing, and produce materials at the end of processing. Materials are defined using the Elements panel in the Definitions window and have an *Initial Quantity* and a *Bill of Material* that defines a list materials and quantities required to produce each unit of the material (e.g. a material named Bike might have a bill that includes 2 Wheels, a Frame, a Fork, etc.). The material that is consumed or produced can be specified as either a single *Material* (e.g. Bike), or as the *Bill of Materials* (e.g. the bill for Bike).

In some cases it is desirable to not start a production lot unless the entire lot can be produced within a specified makespan. For example it may not be desirable to start a lengthy operation immediately before shutting down for the weekend. If you specify the *Maximum Makespan* then the estimated makespan plus the *Makespan Buffer Time* must be less than this *Maximum Makespan*. Note that the *Makespan Buffer Time* is a safety factor to account for unexpected delays.

The estimated makespan is computed based on the shift pattern of the workstation, but does not take into account shift patterns for secondary resources or material shortages. These are fully accounted for, however, in the actual makespan for the production lot. Hence it is possible that a production will start based on its estimated makespan but end up violating the maximum due to constraints.

Worker

The Standard Library has a transporter definition named **Worker** that defines a moveable resource that may be seized and released for tasks as well as used to transport entities between nodes. A Worker object is typically used to model operators, doctors, technicians, etc. To use the Worker you drag a Worker object from the library and place it anywhere in the facility model. You then click on this Worker object to edit its properties.

In contrast to the Vehicle that supports a *Routing Type* that can be *On Demand* or *Fixed Route*, the Worker is assumed to always operate in an on demand mode – i.e. the Worker always waits for either a visit request or a transport request. The Worker always assigns priority to seize visit requests over transport requests. Both the Vehicle and the Worker have the ability to follow a work schedule. When following a work schedule an *Off Shift Action* can be specified as Go *To Home*, *Park at Home*, *Part at Node*, or *Remain in Place*.

Example: Moveable Operators

In this example we have two part types that are produced on separate machines that share a single operator. PartA has an exponentially distributed inter-arrival time with a mean of 3 minutes, and a service time in minutes given by a triangular distribution with a minimum of 1 minute, mode of 2 minutes, and a maximum of 3 minutes. PartB has an exponentially distributed inter-arrival time with a mean of 4 minutes, and a service time in minutes given by a triangular distribution with a minimum of 1 minute, mode of 2 minutes, and a maximum of 3 minutes. The operator is required for processing on either machine, and travels at a rate of .5 meters per second over the 10 meter distance between the machines.

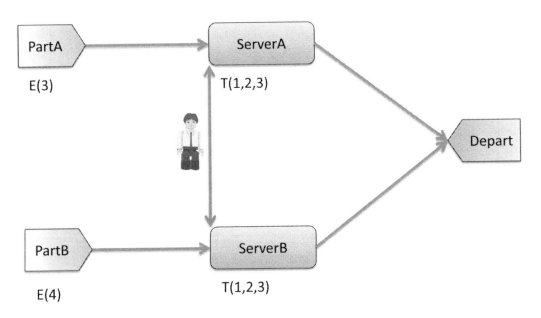

Begin by opening Simio and selecting a new model Project. Now place two Servers (approximate 20 meters apart), two Sources, and a Sink and connect them by paths. Also place two entities from the Project Library and rename them *PartA* and *PartB*, and recolor *PartB* to be red. Change the mean for the *Interarrival Time* on for each of the two Sources, and change the *Entity Type* for *Source2* to *PartB*. Place a BasicNode (default name BasicNode1) near *ServerA* and place a second BasicNode (default name BasicNode2) near *ServerB*. Draw a Path between these nodes and size it to 10 meters, or set Drawn to Scale to false and set its logical length to 10 meters. Change this path *Type* to be *Bidirectional*.

Now drag an object from the Worker definition in the Standard Library and place it in your facility model and name it *Operator*. Specify the idle action for the Operator as *Remain In Place*. Select a symbol from the Project Symbols in the Symbols ribbon to represent the *Operator*. Select *ServerA* and change the parameters for the processing time, and then open *Secondary Resources and Resource for Processing.*, Select the resource Object Name *Operator*, and select the *To Node* option for *Request Visit*, and specify the *Node Name* as *BasicNode1*. This will cause *Operator* to travel to *BasicNode1* before processing will start at ServerA. Repeat these same steps for *ServerB*, except specify the *Node Name* as *BasicNode2*. The model is now ready to run and is shown below.

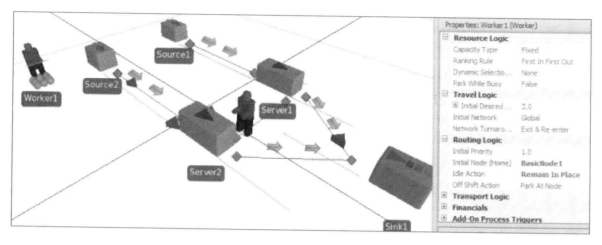

Summary
The Standard Library provides a generic set of object definitions for modeling a wide range of applications. As we will discuss in Chapter 6 you can easily build your own library objects to use in place of the Standard Library. However you can also extend the behavior of the Standard Library definitions using add-on processes. This is a powerful capability that dramatically expands the flexibility and use of this Library. The use of add-on processes is the focus of our next chapter.

Chapter 7: Data Driven Models

Overview

Simulation models typically have data that is used to drive the model. In some cases this data is conveniently entered directly into the modeling objects. For example with simple flow line making a single part it might be convenient to "hard code" the inter-arrival time and processing times directly into the properties for the objects in the model. However in many cases there is a large amount of data as well as a desire to frequently change the data. In addition the people using the model with different sets of data may not be the same person that builds the model. In these cases it is convenient to be able to define the data in data tables and then have the model reference the data in those tables. It may also be desirable to either import or directly bind the data tables to external data sources such as Excel or a database. This chapter describes the features in Simio for building data-driven models using data tables that interface to external data sources.

Data Tables

A Simio data table is a data container that has a collection of user-defined columns with corresponding rows of data. Each column can hold a specific type of data such as a string, real number, Boolean, date-time, expression value, entity type, etc. Within Simio you can define as many tables as you like, and each table can have any number of columns of different types. You can create relations between tables such that an entry in one table can reference the data held by another table. For example a table of production orders could reference a table of product descriptions.

A data table is defined using the Tables panel in the Data window. To add a new table you click on either Data Table or Sequence Table in the Tables section of the Table ribbon. The sequence table is data table with a column named *Sequence* that has already been added to the table for specifying a routing sequence for an entity. Hence everything that applies to a data table also applies to sequence tables. Once you have added a table you can rename it and give it a description by clicking on the tab for the table and then setting the table properties in the Property window.

If you add multiple tables you can drag the tabs apart and drop them on the layout targets to arrange them as you wish. You can also arrange them into vertical and horizontal tab groups by right clicking any table tab and selecting the appropriate option.

To add columns to a table you select a table to make it active and then click on property types under Standard Property, Element Reference, Object Reference, or Foreign Key. You can then rename the column by selecting the column header and changing the column properties. An element reference is a reference to a Simio element such as a Station or Material. An object reference is a reference to an object instance such as Worker or Lathe. The standard properties are all the other property types in Simio and include the following:

Property Type	Description
Real	A real constant: e.g. 132.7
Integer	An integer constant: e.g. 34 or -27
Boolean	A true/false check box.
Expression	A valid Simio expression involving one or more variables: e.g. *(X + Y) / 2.3*
Date Time	A date time value with a calendar date selector: e.g. *1/1/2010 12:00:00 AM*
String	A valid string: e.g. Fred
Event	An event that is defined in the context of the model: e.g. *ModelEntity.Transferred*
List	A value selected from a list that is referenced by the property: e.g. *Red*
Enumeration	A value from an enumeration that is referenced by the property: e.g. *FirstInFirstOut*
Rate Table	A rate table that is defined for this model: e.g. *ArrivalPattern*
Table	A data table that is defined for this model: e.g. *ProductTable*
Sequence Table	A sequence table that is defined for this model: e.g. *RoutingTable*
State	A state that is defined for this model: e.g. *ReworkCount*
Schedule	A schedule that is defined for this model: e.g. *OverTimeShift*
Day Pattern	A day pattern that is defined for this model: e.g. *StandardDay*
Selection Rule	A dynamic selection rule: e.g. *Smallest Value First*

Note that some properties have required values that must be set in the Properties Editor for the column to work. For example a List property requires that a list specified that defines the valid options for the column. Hence we create a list named *Colors* with value members *Red, Green,* and *Blue*, and then add a List column referencing the *Colors* list; the valid entries for the column are *Red, Green,* and *Blue*.

The columns that specify numeric values (e.g. Real, Integer, Expression) also have a *Unit Type* property that can be specified. The unit type can be *Time, Travel Rate,* or *Length*, and then the appropriate units can be specified (e.g. Meters, Feet per Minute, etc.).

You can also select a column and make it a Key Column. A key column must have all unique entries in the column. For example a key column of type *Integer* cannot have two rows with the same number. A key column can then be referenced by adding a *Foreign Key* column property in another table, and referencing this table and key column in the properties of the foreign key. This will then link the rows in that table to a row in this table using the unique key value in this column. We will see an example of this later in this chapter.

The following shows an example table named MyTable with four columns named W, X, Y, Z. Column W is a real column with *Unit Type* specified as *Time* and units in *Hours*. Column X is a Boolean column where a check mark indicates True. A Boolean True in Simio has a numeric value of 1. Column Y is an entity reference column and is set as a key column as indicated by the key icon in the header. As a result the entries in this column must be unique. Column Z is an expression column and accepts any valid Simio expression.

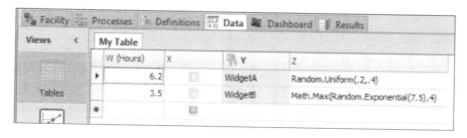

Once a table is defined you can populate the table with data in several different ways. You can manually populate the table by directly entering data for each cell in the table. Note that the bottom row with the * is a row that has not yet been added – once you enter a value in this row it then becomes an active row being edited as indicated by the pen that replaces the '*'. You can also copy and paste within this table, between different tables, and between external data sources (e.g. Excel) and the data table. You can also import from or bind to external data sources.

When working with Excel as an external data sources it is often convenient to first define the columns in Simio, and then export the empty table to Excel, and then populate the data inside Excel. You can then import or bind to the spreadsheet. Importing is a manual operation that happens each time you click on import. Binding is an automatic operation; once the binding is established the data is automatically imported at the beginning of each run with no manual intervention.

You can reference a value in the table using the syntax TableName[RowNumber].ColumnName, where RowNumber can be an expression specifying the 1-based row number. For example *MyTable[1].X* would reference the value *6.2*, and *MyTable[1].Z* would reference the expression *Random.Uniform(.2,.4)*.

It is also possible to have either a token or an object reference a specific row in a table using the SetRow step. Once a row has been assigned to the token/object the values in the table can be referenced without appending the [RowNumber]. For example the MyTable.X references the Boolean X value for the row assigned to the token/Object. If a row has been assigned to a token and then that token is used to create a new object that newly created object will reference the same row as the token.

Example: Multiple Entities from a Single Source

In this example we are going to use a table for generating multiple customer types from a single Source. We have three different types of customers that wait in a single line to be processed by two identical servers. The customers arrive according to a Poisson process, with a mean time between arrivals that is exponentially distributed with a mean of 3 minutes. The customer mix (percentage) and the processing time on the server are summarized in the following table.

Customer Type	Mix (Percentage)	Processing Time
Standard	60	Random.Triangular(3,4,5)
Simple	15	Random.Triangular(1,2,3)
Complex	25	Random.Triangular(7,9,12)

To model this we will place a Source, Server, and Sink into our facility and connect them with Paths with *Allow Passing* set to *False*. We will specify the time between arrivals for the Source as an exponential sample with a mean of 3 minutes, and we will set *Input Buffer* on the Server to 0, and set the *Capacity* to 2. The combination of no input buffer on the Server and no passing on the Path will cause our customers to queue up along the path in front of the Server. The capacity of 2 on the Server will allow this single Server object to represent two identical servers. We also place three instances of

ModelEntity and rename them *Standard*, *Simple*, and *Complex*, and select people symbols from the symbol library to represent these three customer types.

We will now build a Simio table by selecting the Data window and clicking on Add Data Table and then renaming the table *CustomerData*. Next we will add a column named *CustomerType* by clicking on the Object Reference property and selecting Entity. Next we add a column named *Mix* by clicking on Standard Property and selecting *Real*. We then add a column named *ProcessTime* by clicking on Standard Property and selecting *Expression*. We also set the *Unit Type* for this property to *Time* and set the units to *Minutes*. We then manually enter the data for the table as shown below:

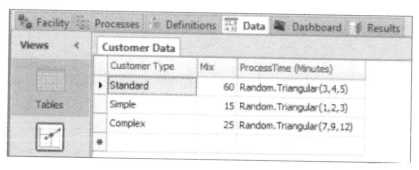

Now that we have this table defined we will use it to select the type of entity to create at the Source, and specify the processing time at the Server. To do this we will specify a table reference assignment for our Source object *Before Creating Entities*. We will specify the *Table Name* as *CustomerData*, and specify the *Row Number* as *CustomerData.Mix.RandomRow*. Note that *RandomRow* is a function that returns a random row number based on the values in the *Mix* column as weights. We can specify the *Entity Type* as *CustomerData.CustomerType*. The newly created entity will then point to this same row. We then change the processing time for the Server to be *CustomerData.ProcessTime*. Hence the processing time will be taken from the *ProcessTime* column for the selected row.

The running model for this example is shown below. Note that you may need to lengthen the Processing.Contents animation queue to see both customers in service at the same time.

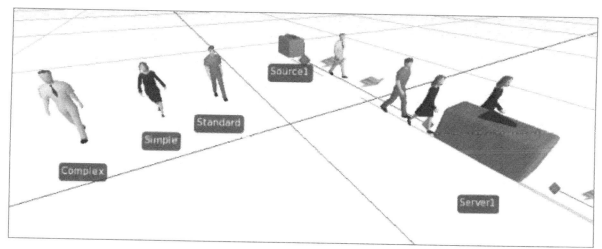

This example is simple but illustrates the power of tables for building data driven models. Note that we could also bind this table to an Excel spreadsheet and then do all of our data editing in the spreadsheet.

Example: Product Routings using Separate Sequence Tables

In this example we have three part types with each part type having a different routing through a facility comprised of three servers. The time between part arrivals is exponential with a mean of 1 minute, and the product mix for the three part types is 20% type A, 30% type B, and 50% type C. The routing and processing time for each of the three part types is summarized in the following tables:

PartA

Station	Process Time
Server1	Uniform(.5,.9)
Server2	Triangular(.5,1.1,1.2)
Sink	

PartB

Station	Process Time
Server3	Triangular(.5,.8,1.2)
Server2	1.5
Server1	1

PartC

Station	Process Time
Server2	Triangular(.5,1.2,1.6)
Sink	

To model this we will place a Source, three Servers, and Sink, and connect them with a network of paths to allow for movements in all directions as shown below. We set the routing logic for all of the output nodes to *By Sequence*. We have also placed three instances of ModelEntity (PartA, PartB, and PartC).

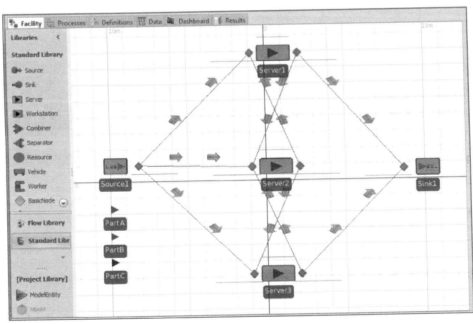

Next we will build our data tables. In this example we are going to use separate sequence tables for each part type, but in the next example we will redo this same problem using data relations to store all three sequences in a single sequence table. We will add data table named *JobTable*, and sequence tables named *SequenceA*, *SequenceB*, and *SequenceC*. For the sequence tables we will set the *Accepts Any Node* property on the Destination column to False so that we can use the abbreviated notation in the routing (i.e. Server1 instead of Input@Server1). We will add one additional column to each sequence table named *ProcessTime*.

For the *JobTable* we will add an Object Reference property of type Entity named *PartType*, and Expression property named *ProcessTime*, a real column named *ProductMix*, and a Sequence Table property named *PartSequence*. The purpose of the *ProcessTime* column is to create an "indirection" to allow us to reference the process time in our model using *JobTable.ProcessTime*, regardless with sequence table is holding the processing time. In a similar way we can reference the part sequence as *JobTable.PartSequence* independent of which part type we have. The following shows the *JobTable* and three sequence tables populated with the appropriate data. Note that the tabs of the three sequence tables have been dragged to display them below the job table.

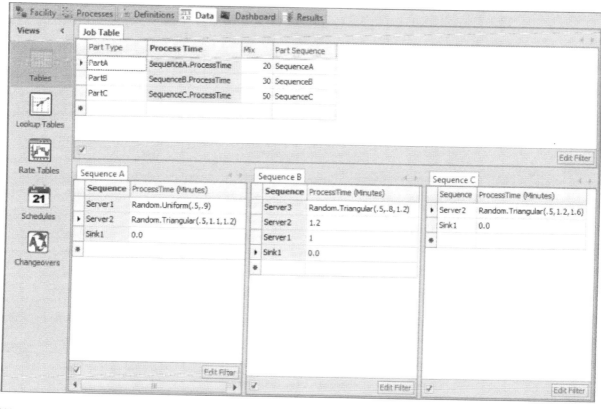

We now specify **table reference assignments** at our Source. The first is *Before Creating Entities* (just before entity creation) and sets the active row for the token to a random row in *JobTable* based on the *Mix* column. The second table reference assignment is *On Created Entity* (just after entity creation). Here we have the entity point to the sequence table specified in the *PartSequence* column in *JobTable* by specifying the table name as *JobTable.PartSequence*, and initial row number as 1 (the first row in the sequence table). We can now specify our *Entity Type* on the Source as *JobTable.PartType*, and our processing time on each Server as *JobTable.ProcessTime*. Note that the value returned is retrieved from the active row of the appropriate sequence table for that part type. The following shows our model running.

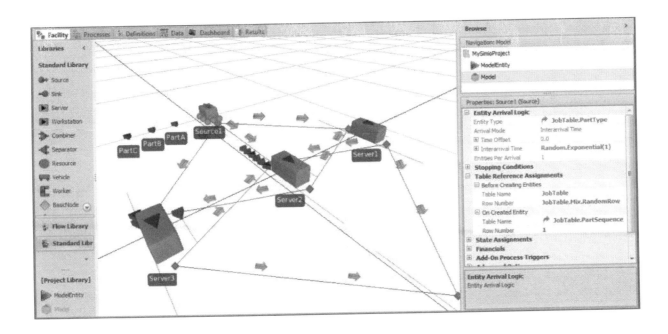

Example: Product Routings using a Table Relation

We will now redo this same problem using a foreign key to create a table relation. This approach allows us to exploit relational data design to simplify our tables. We keep the same facility model as before, except we will be changing a few table references per our new table design.

In our new approach we will have only two tables: one to hold the job data and one to hold the sequence steps for all three part types. Our basic table design is shown below. The *Job Table* specifies the *Part Type*, *Product Mix*, and the *Sequence Type* to follow. The *Sequence Type* is set as a key column and hence all values in the column must be unique. The *Sequences* table specifies the *Sequence*, *Sequence Type*, and *Process Time*. The *Sequence Type* is a Foreign Key property that references the *Sequence Type* column in the Job Table. Note that multiple rows in our Sequences table reference the same row in the Job Table.

Job Table

Part Type	Product Mix	Sequence Type
PartA	20	A
PartB	30	B
PartC	50	C

Sequences

Sequence	Sequence Type	Process Time
Server1	A	Random.Uniform(.5, .9)
Server2	A	Random.Triangular(.5, 1.1, 1.2)
Sink1	A	0.0
Server3	B	Random.Triangular(.5, .8, 1.2)
Server2	B	1.5
		1
Sink1	B	0.0

We build our tables in Simio using Tables panel in the Data window, first creating Job Table, and then creating our Sequences table. We define the Sequence Type in the Job Table as a string column, and then select this column and click on Set Column As Key to make this a key column. We then add our Sequences table, and then click on Foreign Key to add a foreign key property named *SequenceType*, and specify the Table Key for this foreign key property to *JobTable.SequenceType*. The following shows the completed and populated tables. Note that the sequences for each part can be expanded/contracted by clicking on the +/- box to the left of the part name.

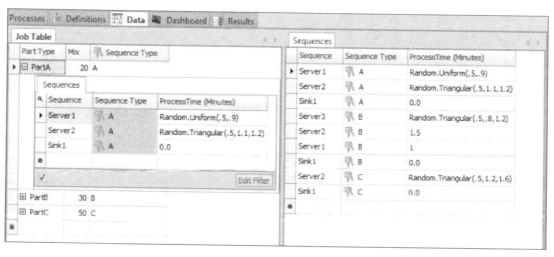

With this data schema we only need to have the entity reference a row in *JobTable*, and then it automatically has access to the corresponding sequence data in the Sequences table through the relation that has been established with the foreign key reference. For example setting a token to point to row 2 in the *JobTable* also establishes *Server3-Server2-Server1-Sink* as the routing sequence. Within the Server we reference the processing time as *Sequences.ProcessTime*, and the value is automatically selected based on both the part type and its current sequence step within its routing. For example *PartB* will use *1.5* minutes for its processing time at *Server2*.

This relational data schema has a number of advantages over our previous approach of creating a separate table for each part routing. Besides having fewer tables and indirections we can also add additional routing sequences to our model without changing the data schema. This is particularly important when importing routings from external data sources.

Summary

This chapter has reviewed the basic concepts for creating data driven models. The table features of Simio can be used to create models that obtain their input data from Excel spreadsheets and other external data sources making it very easy to manipulate the data for the model.

Chapter 8: Processes

Overview

Object-based tools such Simio are very good for rapidly building models. You simply drag objects into the workspace, set the properties for those objects, and your model is ready to run. However the problem that can be encountered with this approach is modeling flexibility. Although the Standard Library can be used to model a wide range of systems, it difficult to have a set of objects that work perfectly in all situations across multiple and disparate application areas without making the objects overly complicated and difficult to learn and use.

The Simio Standard Library addresses this problem through the concept of add-on processes. This advanced functionality is only available with the Design and Team Editions of the product. An add-on process is a small piece of logic that can be inserted into the Standard Library objects at selected points to perform some custom logic. This custom logic can be used to seize/release resources, make assignments to variables, change travel networks, evaluate alternatives, etc. The processes are created as graphical flowcharts without the need for programming. Hence Simio Design/Team Edition combines the benefits of object-based modeling (i.e. ease of learning and rapid modeling) with the power and flexibility of graphical process modeling.

Process logic can be inserted into an object on an instance by instance basis, without modifying or changing the main object definition. For example one Server instance might incorporate process logic to seize and move a secondary resource during processing, while another instance for the same Server definition incorporates special logic to calculate the processing time based on a learning curve, and a third instance of the same Server incorporates special process logic for modeling a complex repair process. All these instances share the same Server definition but customize each instance as needed.

Another thing that is particularly powerful about add-on processes is that – unlike a programming insert - processes can span simulated time. For example a process can wait for tank to fill, a resource to become idle, or a queue to reduce to a specific size. Hence processes are not only easier to learn, create, and understand, they are also significantly more powerful than programming inserts.

Processes

A **process** is a sequence of **steps** that is executed by a **token** and may change the state of one or more **elements**. As the token moves through the process it executes the actions as specified by the step. A single process may have many active tokens moving thru it in parallel. For example the following simple process flow waits to seize a resource, delays by time, and then releases the resource. There may be multiple tokens waiting in the queue to seize the resource, as well as multiple tokens in the Delay step (assuming the resource has a capacity > 1). Note that each step will have properties that govern its behavior – e.g. the Delay step has a property that specifies its delay time.

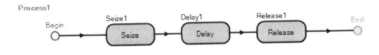

A process always executes within the context of a parent object. When you create a new process in your model the parent object is your model. The properties that are specified for each step may reference properties and states that are "visible" within that context. These properties and states include those that are owned by the model, as well as objects or elements that have been placed into the model. If you have objects that contain other elements or objects in their model definition, then they are also in the context as long as they have been defined with a public scope.

The token that executes a process may also carry its own properties and states. For example a token might have a state that tracks the number of times the token has passed through a specific point in the logic. In many cases you can simply use the default token in Simio, however if you require a token with its own properties and states you can create one in the Token panel within the Definitions window.

A token carries a reference to both its parent object and its associated object (if any). The associated object is the related object (separate from the parent object) that triggered this process to execute. For example a process that is triggered by an entity arriving to an object will have that entity as the associated object. The process logic can reference properties, states, and functions of both the parent object (i.e. the current model) and the associated object. Hence the token could reference the arriving entity to specify the delay time on the Delay step. To reference the associated object it is necessary to precede the property or state name with the object class name. For example *ModelEntity.TimeCreated* would return the value of the *TimeCreated* function for the associated object of type *ModelEntity*. A runtime error occurs if the associated object is not the specified type.

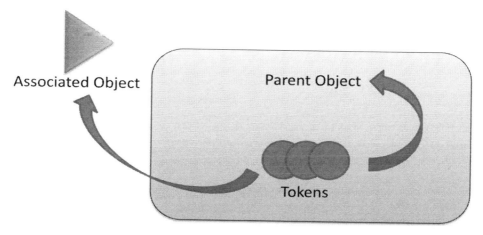

The steps in Simio have properties – but do not have states. Hence steps can have input values that control their behavior, but cannot have variables that change value during the run. The ability to hold changeable state information is provided by Simio elements. Many of the steps in Simio reference a specific element (or object, since an object is sub-classed from element). For example the Consume/Produce steps (which consume and produce material) reference a Material element. Elements are added to your model using the Elements panel in the Definitions window

The token also has a *ReturnValue* that can be assigned during the process. The *ReturnValue* is typically used with special processes called decision processes that are executed in zero simulated time (i.e. no time/status delays – such as a Delay step – are permitted) to make a specific decision. For example a process is executed for a Transporter to see if it is willing to accept a new reservation. For a decision process a *ReturnValue* of 0 denotes *False* and a non-zero value denotes *True*.

Process Types
There are three basic types of processes in Simio. All of these processes are built and edited in exactly the same way and only differ in the mechanism that triggers their execution.

A **standard process** is a process that is explicitly defined by and executed by the Simio engine. If you add a standard process to your model you do not have to specify a triggering event since the Simio engine will automatically execute a standard process at the appropriate time. For example the *OnInitialized* standard process is automatically called for each object by the Simio engine during initialization of the model. You add a standard process to your model by clicking on Select Process icon in the Process window for the model.

An **add-on process** is a process that is incorporated into the model of an object to allow the user of that object to insert their own custom process logic into the model at selected points. The add-on process is automatically executed within the objects model (using the Execute step), and hence there is no need to specify a triggering event for the process. The Standard Library makes extensive use of add-on processes. For example, the Server object provides the following list of add-on processes.

Add-On Process	Logical Point in Server Execution
Run Initialized	During initialization of the model.
Run Ending	When the simulation run is ending.
Entered	Immediately after an entity enters the Server and is about to start the Transfer-In Time.
Before Processing	When the entity has been allocated server capacity but before entering the processing station.
Processing	Just before an entity begins processing.
After Processing	When an entity has completed processing and is about to exit from the processing station to release the server capacity.
Exited	Immediately after an entity exits the Server model.
Failed	Immediately after the Server enters a failed state.
Repaired	Immediately after the Server returns to a repaired state.
On Shift	Immediately after the Server goes on shift.
Off Shift	Immediately after the Server goes off shift.

These add-on processes give considerable flexibility in terms of extending the logic executed by the Server model. For example in the case of the Server the *Failed* add-on process could be used to seize a repairman, and then the *Repaired* add-on process could be used to release the repairman.

There are two ways to insert a new add-on process into the model. The manual way is to click on the Create Process icon in the Process ribbon for the Processes window, and then select that process for one of the Add-On Process Triggers in the object that is placed in the facility model. The quick way is to simply double click on the name of the Add-On Process Trigger, which creates the process with a default name and then selects that process for that trigger.

Note that what you are doing with an add-on process is creating a logical process in the context of your model, and then passing that into an object such as a Server to execute inside that model for that only that one instance of the Server. The same add-on process can be passed into any number of objects.

An **event-triggered process** is a process that is triggered by an event that fires within the model. For example a station element named *Input* (which might be used to define an input station within the object) fires an *Input.Entered* event each time the *Input* station is entered. You can have a process be triggered to execute by this event by specifying the *Triggering Event* property for the process as *Input.Entered*. A token will then execute this process each time an entity enters the station, and the associated object for the token will be the entering entity. It is also possible to have multiple triggering events and to dynamically change the triggering event during the running of the simulation by using the Subscribe and Unsubscribe steps in the process logic.

You create an event-triggered process by clicking on the Create Process icon in the Process tab of the Processes window, and then specify the *Triggering Event* property for that process.

Building Processes

You build and edit processes using the Process window shown below. The left side of this window has a fly-out panel for selecting steps, and the right hand side has the navigation panel and properties editor. The steps are categorized by Common Steps, All Steps, and User Defined Steps. The Common steps include the steps that are most frequently used in models. The All Steps include all of the built-in steps for Simio. The User-Defined Steps include user-coded steps that have been added to Simio to extend the core functionality of the product (for details search for "API" in the Simio help or see *Simio API Reference Guide.chm* in the Simio Program files folder).

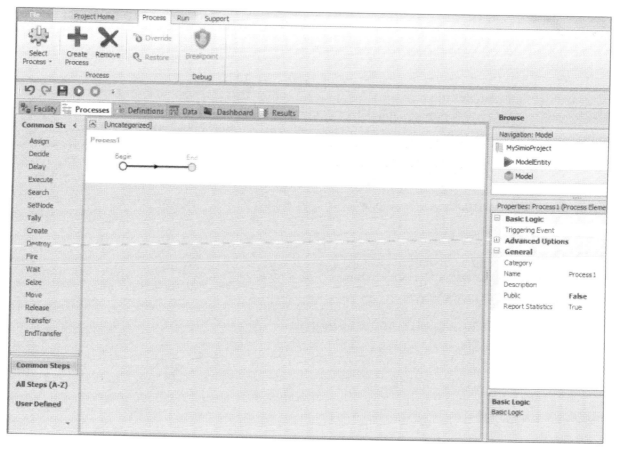

When a new process is added to the window it is shown with a Begin and End point and no steps in between. You add steps to the process by dragging them out from the panel and dropping them at the desired location. Simio incorporates a layout tool that automatically redraws the process as you drop each new step. You can drag a step to move it to a new location within the layout. You can also drag one of the End points to connect it to the input of step. The following shows the result of the automatic layout with multiple Decide steps, where the End coming out of the Delay step has been connected to the input side of the Assign.

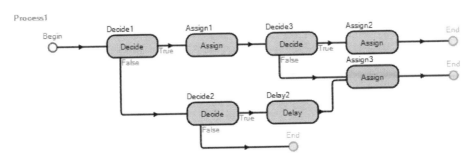

You select a step by clicking on the step, and you select the process by clicking outside the steps. If you select a step the properties of the step are displayed in the Properties window. If change the properties of a step from their default values and then hover over the step a box will appear showing the non-default properties. If you select a process the selected process is shown with slanted hash lines and the properties for the process are displayed in the Properties window. You can zoom the selected process in an out using the + and – keys. If the process drawing is wider the window you can pan left and right by clicking anywhere in the process outside of a step and dragging left and right.

The process properties include the *Triggering Event* (only used for event-triggered processes), and the *Token Class Name* for the token that executes this process (new Token Classes can be defined using the Tokens panel in the Definitions window). The *Initially Enabled* property allows you to set the initial value of the enabled state for the process. The property named *On Associated Object Destroyed* (under Token Actions) specifies what should happen to a token running this process if the associated object that triggered this process happens to get destroyed while this process is running. The property *On Associated Object Transfer Requested* (also under Token Actions) specifies what should happen to a token running this process if the associated object that triggered this process happens to request a transfer to a new location. *End Process* and *Continue Process* are the options for both the token actions. Under *General* you can assign the process to a new or existing *Category*, specify the *Name* for the process, give it a *Description*, define if it has *Public* scope, and specify if the process should *Report Statistics*. Note that all processes assigned to the same category will be grouped together and the groups can be individually collapsed and expanded. You can copy and paste processes. If you paste a process external to Simio (e.g. in a Word document) you can use the Special Paste option to past the process as either a bitmap drawing or as an XML description.

Steps and Associated Elements

The key to process modeling is gaining an understanding of the functionality of the available steps and elements, and how to combine these into process flows to model different aspects of the system. Some of the steps and elements you may use quite often, and others you may never use. One way to approach learning these steps is to get a brief understanding of all the steps, a more detailed understanding of the common steps, and then learn the steps in detail as the occasion arises for the their use. Hence we will proceed with a broad overview of the types of actions supported by the various steps and elements, without including all of the details for the steps (detailed descriptions of the steps and elements may be found in the Simio Reference Guide). We will then give a number of examples of the use of these steps for extending the logic of the Standard Library.

As noted earlier many of the steps in Simio reference elements that hold changeable state information on the system. The elements are added using the Elements panel in the Definitions window. A summary of the Simio elements is shown in the following table:

Element Name	Description
Station	Defines a capacity constrained location within an object.
Timer	Fires a stream of events according to a specified *IntervalType*.
Failure	Defines a failure state for the parent object.
BatchLogic	Forms a batch by matching entities together and attaching them to a parent entity.
Storage	Defines a queue for temporarily storing one or more entities in a specified order.
Monitor	Fires an event when states change or cross a specified value.
Network	Defines a network of links.
RoutingGroup	Defines routing logic for selecting a destination node from a candidate node list.
OutputStatistic	Defines an expression that is recorded at the end of each replication.
StateStatistic	Records time-persistent statistics on a state variable.
TallyStatistic	Holds summary statistics for observational values recorded using a Tally step.
Operation	Defines a sequence of activities that are preformed over time.
Activity	Defines an activity that is performed within an operation.
Material	Defines a material that can be produced or consumed and own a bill of material.
CostCenter	Defines an area of responsibility where costs are incurred or accrued.
Container	Defines a volume/weight capacity constrained location for holding entities representing fluid or other mass.
Regulator	Regulates transfers of entities into or out of locations.

Most of the steps are general purpose steps that can be used by any of the five object types. However some steps are specific to a specific object type such as an entity, transporter, node, or link. We will begin first describing the general purpose steps which are summarized in the following table. Note that the shaded steps are those included in the Common Steps list and are the ones most often used. Where appropriate the elements that are referenced by the steps are also noted. Elements are added to the model using the Elements panel in the Definitions window.

General Purpose Steps

Step Name	Action
Assign	Assigns an expression value to a state variable.
Decide	Sends the token to one of two exit points based on an expression.
Delay	Delays the token for a specified time.
Execute	Executes a process and either waits or continues.
Find	Find the value of one or more indexes in an expression involving arrays.
Search	Search a collection of items including table rows or objects in a list or queue.
SetNode	Sets the current destination for the parent or associated object.
Tally	Tallies a value in the specified TallyStatistic element.
Create	Creates a new entity.
Destroy	Destroys the parent or associated entity.
Fire	Fires the specified object event.
Wait	Waits for a specified event to fire.
Seize	Seizes one or more resource objects.
Release	Releases one or more resource objects.
Move	Requests a move from one or more moveable resources.
Transfer	Transfers the associated entity between objects and/or free space or between stations.
EndTransfer	Completes the transfer of the associated entity into the object and/or station.
AddRow	May be used to add a new row to a specified output table.
Allocate	Manually trigger resource allocation for the parent object.
Batch	Create a batch with the associated entity using the BatchLogic element.
ClearStatistics	May be used to clear model statistics.
Consume	Consume a specified quantity of material.
EndActivity	End the current operation activity for the associated entity.
EndOperation	End the current operation for an entity.
EndRun	Sets the ending time of the simulation to the current simulation time.
Fail	Specifies a Failure element to change to a failed state.
Insert	Insert the associated or parent object into a queue.
Interrupt	Maybe be used to interrupt process delays.
Move	May be used to request a move from one or more moveable resources.
Notify	May be used to output a user-defined trace or warning message.
Produce	Produces a specified quantity of a material.
Remove	Remove the associated or parent object from a queue.
Repair	Specifies a Failure element to restore to a repaired state.
Resume	Resumes a suspended process or movement of the parent or associated object.
SetNetwork	Sets the current network for the parent or associated object.
SetRow	Sets a table reference and row for the token or parent/associated object.
StartActivity	Starts a specified activity for an operation.
StartOperation	Starts a specified operation.
Subscribe	Adds a new triggering event to a process.
Suspend	Suspends a process or movement of the parent or associated object.
Travel	Moves an entity in free space to a specified location.
UnSubscribe	Removes a triggering event for a process.

The **Assign** step is used to assign an expression value to one or more state variables. The Assign step displays a single assignment in the main properties grid, but allows for additional assignments in the Repeating Properties Editor.

In addition to the Assign step, there are several other steps used to set dynamic characteristic of an entity/object including **SetNetwork** (specifies the network to follow), **SetNode** (specifies the destination to travel towards), and **SetRow** (specifies a row in a data table for retrieving data).

The **Decide** step is used to select one of two exit points based on the value of an expression. The expression is interpreted either as a logical condition or a probability based on the *Decide Type* property.

The **Delay** step is used to delay the token by a specified time. This step is typically used to model general processing delays in the system.

The **Execute** step is used to execute another process. This step will either hold the current token until the specified process is completed, or continue and execute in parallel to the spawned process.

The **Find** step is used to examine an expression using involving array variables to find the indexes to match, minimize, or maximize the value of the expression.

The **Tally** step is used to record a value for a specified TallyStatistic element. To use the Tally step you must first add a **TallyStatistic** element to your model. The Tally can record either the value of an expression, or the time between arrivals to the TallyStatistic based on the *Value Type* property specified for the step. Note that *Token.TimeInProcess* can be used in the expression to record the time that the token has spent in the process, and *Entity.TimeInSystem* can be used in the expression to record the time that the associated entity has spent in the system.

The **Create** and **Destroy** steps are used to create and destroy dynamic objects which include entities and transporters. The Create can either create a new dynamic object, or create a copy of either the parent object or associated object. The Destroy step can destroy either the associated object or the parent object (i.e. destroy itself).

The **Fire** and **Wait** steps provide coordination between logical points within an object, or between objects. The Fire step is used to fire an event that has been defined for the object (events are defined using the Events panel in the Definitions window). The Wait step is used to hold an entity until a specified event has been fired. The Wait step also includes a *Release Condition* property that if specified must be true for a token to be released.

The **Seize** and **Release** steps are used to seize and release capacity of an object that has been defined as a resource. As noted in Chapter 3 that Resource object can be used to model fixed resources, and the Vehicle object or an Operator entity can be used to model a moveable resource. Both the Seize and Release steps utilize the Repeat Group Editor to allow for multiple seizes and releases at a single step. Note that in the case of the Seize you can request a visit by a moveable resource using the *Request Visit* property under Advanced options.

The **Transfer** and **EndTransfer** steps are used to initiate and complete entity transfers between objects, between an object and free space, or between stations within an object. At the completion of the transfer the EndTransfer step fires the entity's *Transferred* event. These steps are used to provide a "handshake" between objects in the model. We will discuss the use of these steps in detail in Chapter 9 where we discuss building custom objects.

There are a number of more advanced steps and steps that are specific to use with specific object types – refer to the reference guide or help for more details on those steps.

We will now turn our attention to using a number of these steps in the following examples.

Example: Using a Moveable Resource with Server Failures (Seize/Release)

In Chapter 3 we modeled a simple system with two servers requiring an operator to move back and forth to process a part on the server. In this example we will replace the operator with a repairman that is required to perform repairs on each of the servers. PartA has an exponentially distributed inter-arrival time with a mean of 3 minutes, and a service time in minutes given by a triangular distribution with a minimum of 1 minute, mode of 2 minutes, and a maximum of 3 minutes. PartB has an exponentially distributed inter-arrival time with a mean of 4 minutes, and a service time in minutes given by a triangular distribution with a minimum of 1 minute, mode of 2 minutes, and a maximum of 3 minutes. The repairman travels at 2 meters per second from the repair depot that is 50 meters away. The repairman can also travel directly from one server to the other along a pathway that is 10 meters long. When the repairman completes the last repair he returns to the repair depot where he waits for the next failure. Failures occur randomly with an exponential time between failures and a mean of 100 hours. The time to perform a repair in hours has a triangular distribution with a minimum value of 1, most likely value of 2, and a maximum value of 3.

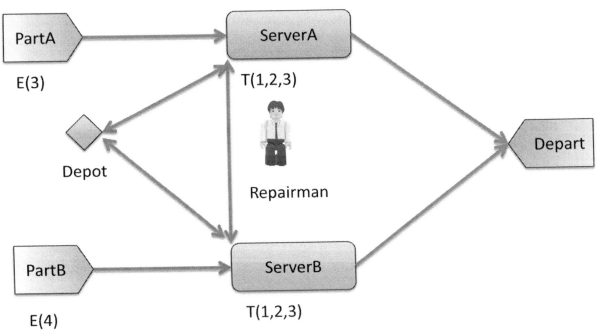

The Server object includes support for a secondary moveable resource during processing; however it does not provide support for a moveable repairman for repair operations. Hence we will use processes to add this missing functionality to the Server object.

We proceed by building our model as before, only omit the Operator, and instead place a Worker named *Repairman*. We then draw the Repairman network as three BasicNodes names *NodeA*, *NodeB*, and *Depot*, connected by bi-directional paths of the specified length (by drawing to scale or setting the

Logical Length). We then set the *Initial Node* for the *Repairman* to the Depot, and set the *Idle Action* to *Go To Home*. We add the missing resource logic to the Servers by inserting add-on process for the triggers named *Failed* and *Repaired*. The *Failed* trigger occurs immediately before the repair activity begins, and the *Repaired* trigger occurs immediately after the repair activity is completed. We will add a Seize step to each *Failed* add-on process, and a Release step to each *Repaired* add-on process, as shown below.

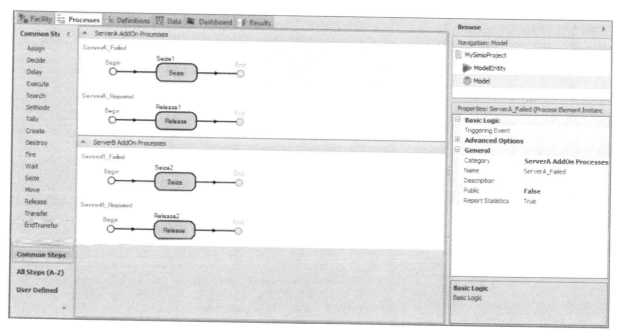

On the Seize steps we specify *Repairman* as the resource to seize, and under the *Advanced Options* we specify a visit request to *NodeA* for *ServerA*, and *NodeB* for *ServerB*. On the Release step we specify *Repairman* as the resource to release. The model is now ready to run and is shown below:

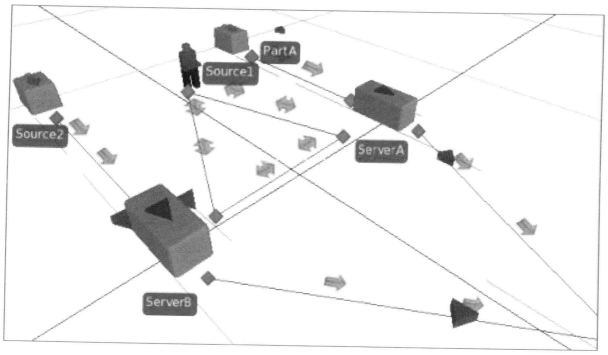

Example: Initializing an Array using a Table

In Chapter 2 we discussed the concept of state arrays. One of the options noted for defining and initializing a state array was the [Table] option. We will now give an example where we initialize a state array with a table, and then make assignments to that array from within the model.

In this example we have four Server locations that are linked together by unidirectional paths as shown below. Parts always enter the system at Server1, and must visit the other three Servers in any order before departing the system. Once they have visited all four Servers they then depart the system. For example a part leaving *Server2* either proceeds to Depart (if all operations are done), or proceeds to *Server3* or *Server4*.

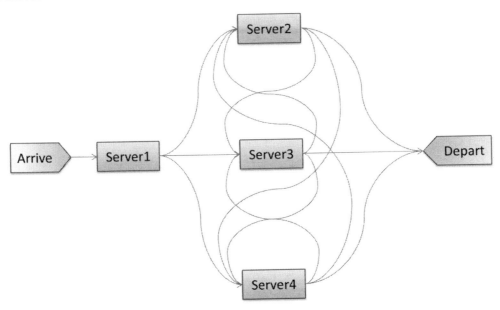

Whenever more than once choice is available for the next Server to visit, we want to choose the path that has been least traveled. For example if we have an option departing *Server2* to go to either *Server3* or *Server4*, we will make this choice based on a count of the number of parts that have travelled each path. We will keep this count in an array that we wish to initialize with an initial count as follows, and then bump by 1 each time we travel the path.

From/To	Server1	Server2	Server3	Server4
Server1	0	12	16	13
Server2	0	0	14	10
Server3	0	17	0	15
Server4	0	9	8	0

We will begin by building our facility model. We place a Source, 4 Servers, and a Sink, and then connect them with paths as shown below:

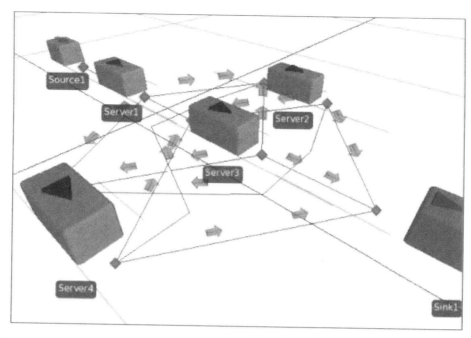

Next we will define our data table for initializing our *PathCount* state array. We add a data table named *PathTable* and create real valued columns named *Server1, Server2, Server3,* and *Server4*, and then populate this table with counts for each travel path as follows:

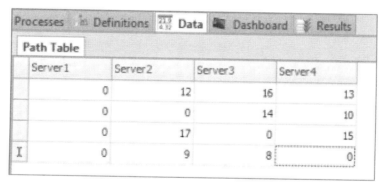

Server1	Server2	Server3	Server4
0	12	16	13
0	0	14	10
0	17	0	15
0	9	8	0

To create our state variable for our main model we click on the States panel in the Data window and then add Discrete State of type Real, and rename our new state *PathCount*. We then specify the dimension as *[Table]*, with the *TableName* specified as *PathTable*. Note that this will automatically dimension our *PathCount* state array and initialize it with the values given in the table. If we were to bind the *PathTable* to an external Excel spreadsheet we could then edit the initial counts directly in Excel.

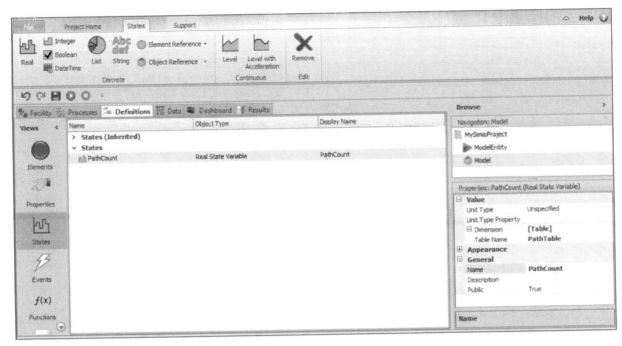

When departing *Server1* we will make our select our next destination based on the smallest current count for the three outgoing paths using the following process logic in the Output node at *Server1*. If the path count from Server1 to *Server2* is less than the current count to both *Server3* and *Server4* we select *Server2*, otherwise we select between *Server3* and *Server4* based on the smallest path count. We enter the expression for Path2 <= Path3&4 as:

PathCount[1,2] <= Math.Min(PathCount[1,3], PathCount[1,4]).

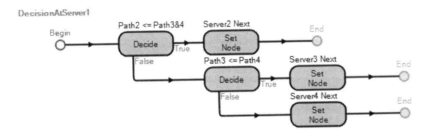

To support the destination selection logic at Servers 2, 3, and 4 we will keep a count on the number of Servers visited by each part, as well as the index of the last Server visited. We select our *ModelEntity* and click on the States panel in its Definition window and add a Discrete State named *LastServer*. For the number of Servers visited we will make use of the existing *Picture* state variable on *ModelEntity*. We will bump this value by 1 each time a the part leaves Server 2, 3, or 4, and then use a Green, Red, Blue, and Yellow entity symbol animate the number of Servers that have been visited by each part.

When departing Server2 we will use the following process logic.

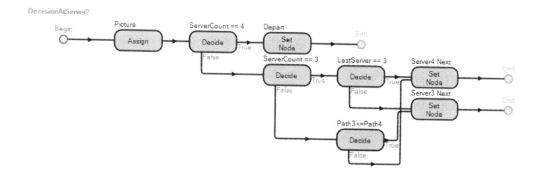

We first increment the Picture state variable and then check if the server count is equal to four. Note that this expression is entered as *ModelEntity.Picture == 3* since we are using the Picture state for a dual purpose: to set the zero-based symbol index, and to track number of servers completed (not counting Server 1). The expression for LastServer == 3 is entered as *ModelEntity.LastServer == 3*.

If the server count for this part is 4 (i.e. *ModelEntity.Picture == 3*) we go straight to Depart. Otherwise if the server count is 3 we select between *Server3* and *Server4* to avoid revisiting the same Server again as our last Server. Otherwise we have yet to visit either *Server3* or *Server4* so we select between them based on the travel path count. Note that this logic will ensure that we visit each Server exactly once. We define similar logic for selecting our next destination at *Server3* and *Server4*.

We add the decision processes to the Entered add-on process for the output node on each of the Servers. We also add a small assignment process to each of the Exited add-on processes on the output nodes for *Server2*, *Server3*, and *Server4* to update *ModelEntity.LastServer* to 2, 3, and 4 respectively. Finally we add a small assignment process for the Entered process trigger to each of the paths connecting the four servers to update the path count. For example the path connecting *Server2* to *Server3* we increment *PathCount[2,3]* by 1. Note we could have alternatively used the State Assignments available to us on the standard objects. Our model is now ready to run.

Summary

The chapter has introduced the concept of process logic for adding new behaviors to our existing modeling objects. The ability to insert process logic into individual instances of an object is an extremely powerful capability in Simio. In Chapter 6 we will expand further on these concepts when we discuss process concepts for building object libraries.

Chapter 9: Object Definitions

Overview

The Standard Library that we have been using up to this point is just one of many possible libraries that you can use to build your models. Using Simio Design or Team Edition you can build libraries of objects for your own purposes, or build libraries to share across your enterprise. You can also build commercial libraries that you market and sell to others. In this chapter we will examine the basic concepts and methods for building your own object definitions, and sharing them with libraries.

One of the basic principles of Simio is the notion that any model can be an object definition. Models are used to define the basic behavior of an object. It is very easy to take a model that you have developed and then "package" it up for someone else to use as a building block for their models. It is also very easy to build sub-models within your own project, and then build your model using these sub-models as building blocks.

The material we will be discussing here is an extension of the ideas we presented in Chapter 8 on process modeling. In that chapter we learned how to use processes to extend the basic functionality of a specific object instance, without altering the core behavior of the object definition. In this chapter our focus will be on either modifying or extending the core behavior of an object definition - thereby changing the behavior of all instances of that object – or on building new objects from scratch.

Basic Concepts

An object definition has five primary components: **properties, states, events, external view,** and **logic**. The properties, states, events, and external view are all defined in the Definitions window for a model. The logic is defined in the Facility and Processes window for the model.

We began using model properties back in Chapter 4 to create properties that were referenced in our model and could be used for experimentation. These same properties are used to define inputs for a model when used as an object definition. Model properties provide inputs to model logic and remain static during the run. States are used to define the current state of the object. Anything that can change during the execution of the simulation is represented in the object as a state. Events are logical occurrences at an instant in time such as an entity entering a station or departing a node. Objects may define and fire their own events to notify other objects that something of interest has happened. The external view is a graphical representation for instances of the object. The external view is what someone sees when they place the object in their model. The logic for the object is simply a model that defines how the object reacts to events. The logic gives the object its behavior. This logic can be built hierarchically using existing objects, or defined with graphical process flows.

External View

The external view is a graphical representation for instances of a fixed or dynamic object. The external view is not used for nodes or links. The external view is defined by clicking on the External panel in the Definitions window for a model as shown below:

Introduction to Simio

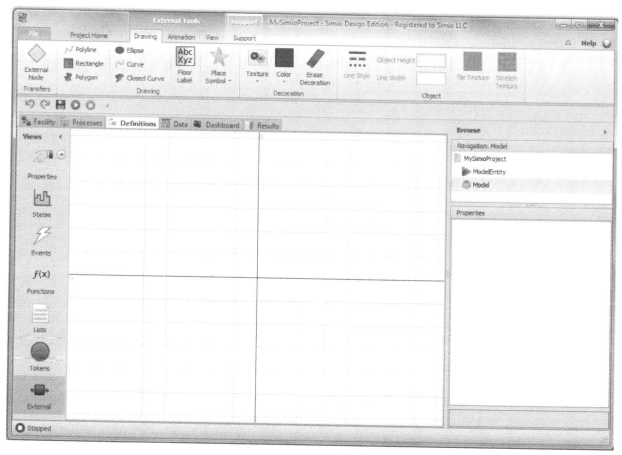

The external view is what a user sees when they place an instance of the object in their model. The graphics for the external view can be inherited from the facility view of the model, or be custom created using symbols from the symbol library or downloaded from Google Warehouse, as well as static graphics drawn using the drawing tools on the Drawing ribbon. In addition the view may contain attached animation components that are placed into the view from the Animation window. These animation components include animated queues, status labels, plots, pie charts, linear/circular gauges, and buttons. Note that in the case of dynamic objects (i.e. entities) these animated objects are carried by each dynamic object in the system. For example an entity could carry a status label showing the value of one of its states, or a button that when clicked by the user causes some action to take place.

When you build a new fixed object you typically provide associated input/output nodes for the object so that dynamic entities may enter and/or leave your fixed object. For example a Server object has associated node objects for Input and output. The characteristics of these associated objects are also defined in the external view of the object by placing external node symbols in the external view using the Drawing ribbon. These external node symbols may also be added to the external view directly from the facility view by using a right-click short cut on the corresponding nodes in the facility view. Note that these are not node objects but symbols identifying the location where associated node objects will be created when this object is placed.

When you place a node symbol in the external view you must also define its properties. The *Node Class* specifies the type of node that should be created. The drop list will give you a choice from all available node definitions which will include *BasicNode* (simple intersection), *TransferNode* (intersection with support for setting destination and selecting transporters to ride), and *FlowNode* (intersection supporting flow type entities), as well as any other node definitions that you have loaded as libraries or are defined in your project. In the case of the Standard Library the *BasicNode* class is always used for input nodes, and the *TransferNode* is used for output nodes. The *Input Logic Type* specifies how arriving entities attempting to enter this object are to be handled by this object. The *None* option specifies that entities are not permitted to enter. The *ProcessStation* option specifies that the arriving entity may enter at a specified station, which fires a station entered event that can be used to trigger process logic. This is used when defining the object logic using a process model. The *FacilityNode* option specifies that the arriving entity is to be sent to a node inside the facility model for the object. This option is used for defining object logic using a facility model. The *ProcessContainer* option specifies that the arriving entity will be entering a specified container defined by the *Container* property. Under *Advanced Options*, *Initial Property Values* may be passed in for instances of the node. In the *General* section the *Name* for the node symbol is specified. This node symbol name is used to create the name of the associated node object that is created at this symbol location when this object is instantiated using the format *NodeSymbolName@ObjectName*. For example if the symbol name is *Input* and the object name is *Lathe*, the associated object that is automatically created is named *Input@Lathe*. In the Standard Library the input node symbols are named *Input* and output node symbols are named *Output*.

Model Logic

The logic for a fixed object definition is defined using a facility and/or process model, as determined by the *Input Logic Type* on the node symbols. In the case of entities, transporters, nodes and links there are no node symbols and hence these models are typically built using process logic.

Whenever you have built a model of a system it represents the logic component of an object definition. You can typically turn any model into a useable object definition by simply adding some properties to supply inputs to the model, along with an external view to provide a graphical representation for the object along with nodes that allow entities to enter and exit the model.

There are three approaches to defining the model logic for an object. The first approach is to create your model hierarchically using a facility model. This approach can also be combined with the use of add-on processes to define custom behavior within the facility objects. This approach is typically used for building up higher level facility components such a work center comprised of two machines, a worker, and tooling. The second and most flexible approach is to create the object definition behavior from scratch using a process model. This is the approach that was used to create the Standard Library. The third approach is to sub-class an existing object definition and then change/extend its behavior using processes. This approach is typically used when there is an existing object that has behavior similar to the desired object, and can be "tweaked" in terms of property names, descriptions, and behavior to meet the needs of the new object.

The concept of sub-classing an object definition is analogous to the concept of sub-classing objects in a programming language. The basic idea is that the new object definition that you are building inherits its properties, states, events, and logic from an existing object definition. The sub-classed object will initially have the same properties and behavior of the original object, and if the original object definition is updated, then the sub-classed object will inherit the new behavior. However after sub-classing you can then hide or rename the inherited properties, selectively override portions of the process logic that is inherited, or add additional processes to extend the behavior. For example you might create a new object definition named *MRI* that is sub-classed from Server and incorporated into a library for modeling healthcare systems. You might hide some of the normal Server properties (e.g. *Capacity Type*), and rename others (e.g. rename *Process Time* to *Treatment Time*). You might also replace the normal internal process that models a failure by a new process that models the failure pattern for an *MRI*, while continuing to inherit the other processes from the Server.

To sub-class an object definition, select the Project in the navigation window, and then select the Models panel.

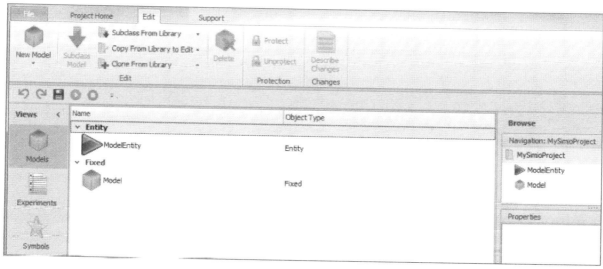

The Edit ribbon lets you add a new blank model, create a sub-class of the selected model, or create a sub-class from an object definition from a library. You can also sub-class a library object using the right-click menu on the library object definition from within the Facility window. You can also create a copy of a library object without sub-classing. Note that in this case the newly created object definition is copy of the library object, but does not inherit its behavior. Hence if the original object definition is changed the copy is not affected. In contrast with a sub-classed object definition changing the original will change the behavior of the sub-classed object since it inherits process logic from its base object definition. You can also protect an object definition with a password. In this case the password is required to view or change the internal model for the object.

The model properties include the *Model Name*, *Object Type* (fixed, link, node, entity, or transporter), *Parent Class* (from which this object was sub-classed), *Icon* for displaying the model, along with the *Author*, *Version* number, and *Description*. The properties also include two flags for specifying if the object has capacity that can be seized and released as a resource, and if the object can be run as a model or only used as a sub-model within other models.

Properties, States, and Events

Whenever you create an object definition you inherit properties, states, and events from the base object class and you can add new members as well. You can view the inherited properties, states, and events in the Definitions window for the model. The inherited and new members are placed in separate categories that can be independently expanded and collapsed. The following shows the Properties panel for a model type Fixed with a single new property named *ReworkTime*. Note that this property can be referenced by both the objects and process flows within the model.

Introduction to Simio

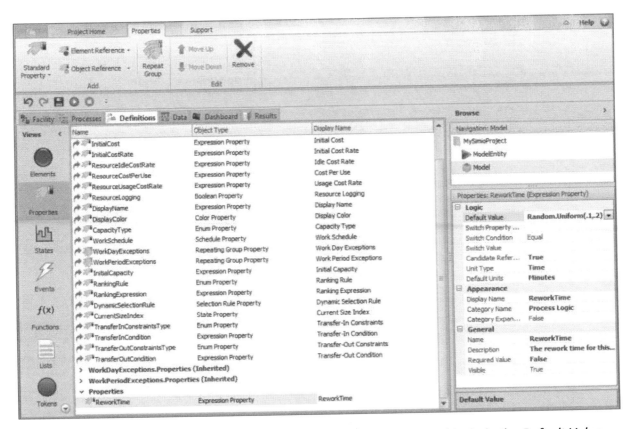

The characteristics of *ReworkTime* are shown in the Property window, and include the *Default Value, Unit Type/Default Units, Display Name, Description, Category Name*, and flags specifying if this is a *Required Value* and if this property is *Visible* to the user. The characteristics also include a *Switch Property Name, Switch Condition,* and *Switch Value*. These can be used to hide or show a property based on the value the user specifies for another property. For example you might have property Y show only if the user sets property X to value greater than 0. This lets you dynamically configure the Property window based on user inputs.

In the case of an inherited property you can change the *Visible* flag along with the *Default Value, Display Name, Description,* and *Category*. Hence you can hide properties or change their general appearance.

Although you can add additional states and events to a model you cannot rename or hide inherited states and events.

Example: Tandem Server

In this example we are going to build a new object definition for a tandem server comprised of two Standard Library Servers in series, with a Path in between. The first Server has a capacity of one and has no output buffer. The second Server also has a capacity of one and has no input buffer; hence the second Server will block the first Server whenever it is busy. The object has two properties that specify the processing time on each of the two servers. The object also animates the entity in process at each Server, as well as animated pie charts for the resource state of each Server.

Starting with a new project add a new fixed model from the Project Home ribbon and rename this new model *TandemServer*. Select the *TandemServer* as the active model and place two Servers connected by a Path with *Allow Passing* set to *False* in its Facility window. Set the output buffer for Server1 to zero, and the input buffer for Server2 to zero. Now in the Properties panel of the Definitions window add two new expression properties named *ProcessTimeOne* and *ProcessTimeTwo*, and set their Default Value, UnitType/Default Units, Display Name, and Category Name as shown below.

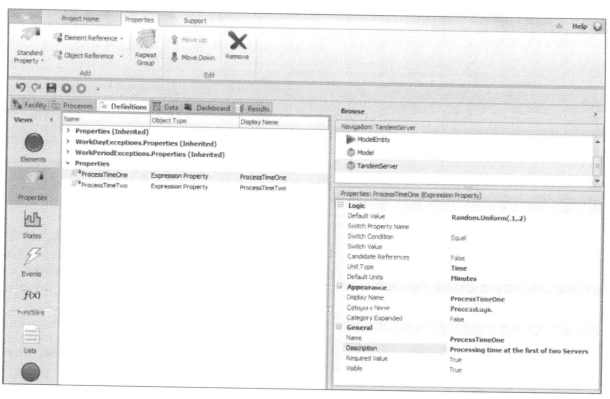

We will now right click on the *Processing Time* for Server1 and select *Set Referenced Property* referencing our newly created property named *ProcessTimeOne*. We do same for *Processing Time* for Server2 referencing *ProcessTimeTwo*. Hence these processing times are specified by our newly created properties for *TandemServer*. Finally we will select each of the two Servers, click on the down arrow next to Status Pie on the Symbols ribbon and select Resource State to draw a pie chart next to each of the two Servers as shown below.

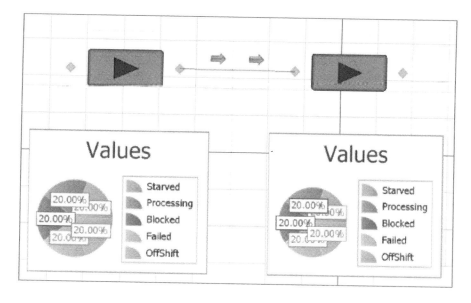

Introduction to Simio

By default the graphics that you place in the Facility view for the Tandem Server will automatically appear in the External view for the object (you can view the current External view by clicking on the External panel in the Definitions window). Hence any user placing the Tandem Server in their model will see the objects in the Facility view. You can selectively hide these objects by right clicking on them and toggling the *Externally Visible* flag to false. You can selectively set this flag for any objects, symbols, or static graphics that you place in your Facility view to completely control the default External view for the object. You can also completely hide the graphics from the Facility view and add custom graphics and animation directly to the External view for the object.

Our final step in building our Tandem Server object is to define the input and output external node symbols for the Tandem Server in the External view. The purpose of these external node symbols is to specify the location and type of associated nodes to create for this object, along with the binding between these associated nodes and the object. Although we can do this step manually, Simio provides a simple automated method for doing this directly within the Facility view. The manual method (which we will skip) is done within the External panel in the Definitions window. Although the manual method works fine, it is far easier to add these symbols to the External view and automatically set their properties by right clicking on the corresponding nodes in the Facility view. We will first right click on the Input node on Server1 in the Facility view and select *Bind to New External Input Node* and specify the name for the external node as *Input*. Next we right click on the Output node at Server2 and select *Bind to New External Output Node*, specifying the name for the external node as *Output*. You can see the results of these actions by selecting the External panel in the Definitions window, and clicking on the two newly added External Node symbols. Note that these external node symbols are placed on top of the associated nodes that are visible from the Facility view.

We are now ready to use our new object definition. Click on our main model (Model) to make it the active model, and then place a Source, TandemServer (selected from the Project Library), and Sink, and connect them with Paths. If you click on the TandemServer you will now see our custom properties displayed in the Property window. The following shows our simple model employing our new TandemServer object in operation.

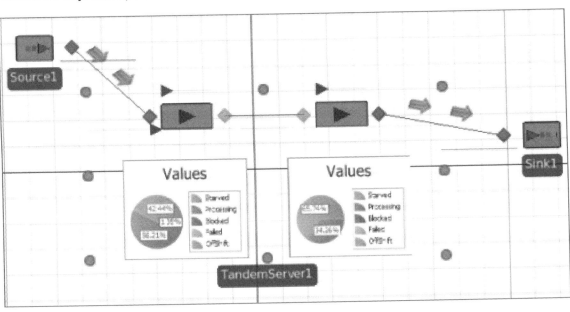

Note that the internal nodes and links that are graphically inherited into the External view of the Tandem Server from its Facility model appear in the External view but are faded to indicate that links in

-91-

the main model cannot be directly connected to these nodes. The entry and exit points into the Tandem Server are defined by associated nodes that we explicitly defined by adding our External Node symbols named Input and Output.

Example: A Base Lathe

In our last example we built a new object definition using a facility model. We will now do the same thing using a process model. Our new object will be a lathe that can process one part at a time. Our lathe has an input buffer for holding waiting parts, and an output buffer for parts waiting to exit to their next location.

Starting with a new project (ModelEntity and Model) add a new fixed model from the Project Home ribbon and rename it *Lathe*. Make *Lathe* the active model and click on the Properties panel in the Definitions window. We will define four new properties for our new object. Within the category named *Process Logic* we will have a *Transfer In Time* (expression, unit type Time, default units minutes, default value 0) for accepting new entities into our input buffer, and a *Processing Time* (expression, unit type Time, default units minutes, default value Random.Triangular(.1,.2,.3)) for processing parts on the lathe. Add two additional properties named *InputBufferCapacity* (with display name *Input Buffer* and default value *Infinity*) and *OutputBufferCapacity* (with display name *Output Buffer* and default value *Infinity*) for specifying the buffer sizes and place them in a new Category Name specified as *Buffer Capacity* by typing in the new Category Name. The following shows our property panel with the attributes displayed for the ProcessingTime property.

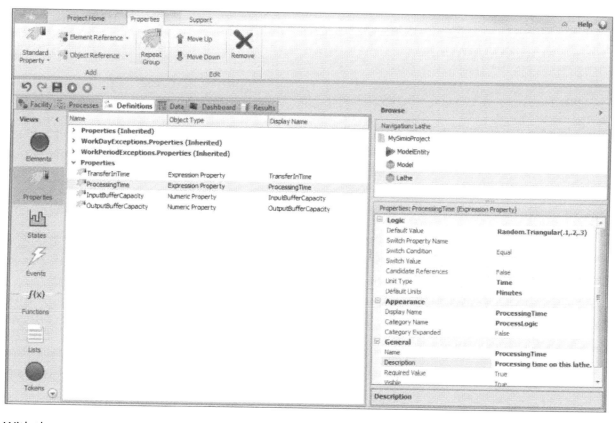

With these properties defined we will then add three station elements named *InputBuffer*, *Processing*, and *OutputBuffer*. Click on the Elements panel in the Definitions window and add and rename these elements. Specify the initial capacity for the *InputBuffer* as our newly created referenced property named *InputBufferCapacity* and the initial capacity for the *OutputBuffer* as our newly created referenced property named *OutputBufferCapacity* (right click on the property, select Set Referenced Property, and select the appropriate referenced property). Specify the initial capacity for *Processing* as 1.

Now we will build three process flows for entering each of these three stations. Entities will enter this object by executing the process flow for the *InputBuffer* shown below:

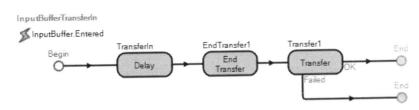

The token that executes this process is triggered by the *InputBuffer.Entered* event and delays by the *TransferInTime* property value before ending the transfer. The EndTransfer step signals to the outside world (perhaps a conveyor, robot, person, etc.) that the transfer is now complete. Once the transfer is complete a new transfer can begin as long as there is remaining space in the *InputBuffer* station. The token then initiates a transfer of the entity from the *InputBuffer* station to the *Processing* station which executes the following logic.

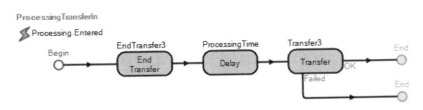

Here the token immediately ends the transfer and then delays by the ProcessingTime property value before initiating a transfer from the Processing station to the OutputBuffer station which executes the following logic.

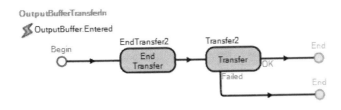

Here the token terminates the transfer and then initiates a transfer from the *OutputBuffer* out through the associated node object that is defined by the *ParentExternalNode* named *Output* (note: we will not be able to specify this node until we have first defined it in the external view).

Next we will define the external view. In this case we have no Facility graphics to inherit so we must use the manual method to define the graphics for the external view, along with the external node symbols for defining the associated input and output nodes. Click on the External panel in the Definitions window and place a lathe symbol from the symbol library in the center of the external view, and add an external node with *Node Class BasicNode* and named *Input* on the left side of the lathe, and an external node with *Node Class TransferNode* and named *Output* on the right side of the lathe (we can now go back and specify our *ParentExternalNode* named *Output* on our *OutputBuffer* Transfer step). Specify the *Input Logic Type* on the *Input* external node as *ProcessStation*, and then select *InputBuffer* from the drop list. Note that arriving entities will transfer to the *InputBuffer* station upon arrival to the associated node object corresponding to this external node symbol. We will add animated queues for animating the InputBuffer, Processing, and OutputBuffer stations. Our external view for the Lathe is shown below:

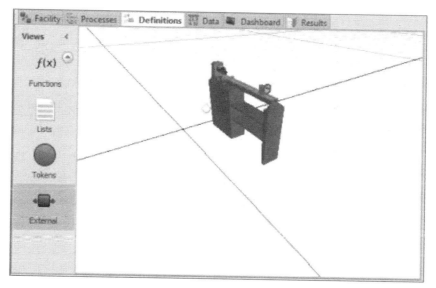

We are now ready to use our new object definition in a model. Click on Model in the navigation window to make it our active model and drag out a Source, Lathe (from the Project Library), and Sink, and connect them with Paths. Click on Lathe to edit its properties. You can now run the model.

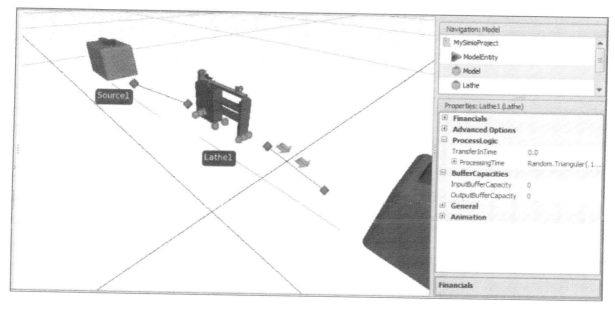

Example: Server with Repairman

In this example we are going to build a new object definition named MRI that is sub-classed from Server. We will hide and rename some properties, and add a new property to specify an optional repair person that is required to perform a repair on the MRI. We will then modify the repair logic to seize and release this repair person.

Beginning with a new project we right click on Server in the Facility window and select Sub-Class (we could do this same operation in the Project window). This adds a new model to our project named MyServer, which we will rename MRI. Click on MRI to make it the active model, and then click on the Properties panel in the Definitions window. Add a new object reference property and rename it RepairPerson and specify its *Category* as Reliability Logic. Specify the *Switch Property Name* as Failure Type, the *Switch Condition* as NotEqual, and the *Switch Value* as NoFailures. Hence this property will not be visible if the NoFailures option is selected by the user. We will also set the *Required Value* flag to False for this property. This property definition is shown below:

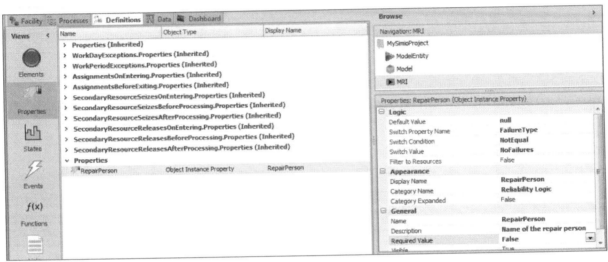

We will also hide the inherited properties named *Capacity Type*, *WorkSchedule*, and *InitialCapacity*, and rename the *Processing Time* property to *Treatment Time* by changing its display name, and enter a new description for this property. From the Properties panel expand the category Properties (Inherited) by clicking on the down arrow on the right side of the category title bar. Hide the *Capacity Type*, *WorkSchedule*, and *InitialCapacity* properties by selecting each property and changing the *Visible* flag to *False*. We also select and edit the display name and description for the *ProcessingTime* property.

Next we will define the external view for the MRI. Click on the External panel in the Definitions window, delete the inherited Server symbol, and then in the Place Symbol drop list select Download Symbol to download an MRI graphic from Google Warehouse and reposition the external node symbols and animated queues for the queue states *InputBuffer.Contents*, *Processing.Contents*, and *OutputBuffer.Contents*. You may also want to change the queue type for Processing.Contents queue to an Oriented Point queue and position it so that patients lay inside the MRI. Note that the two external node symbols are inherited from the Server and can be moved but cannot be modified.

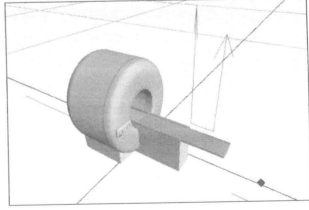

In last step we will modify the process logic that we have inherited to make use of the repair person during maintenance. Click on the Processes window to display the inherited processes from the base Server definition. Note that you cannot edit any of the processes at this point because they are owned by the Server and are inherited for use by the MRI. We will select the first inherited process in MRI named *FailureOccurenceLogic* and then click on Override in the Process ribbon. This will make copy of the inherited process from the Server for use by the MRI in place of the same process that is owned by the Server. We can then edit this overridden process in any way that we wish. We will add a Seize step immediately before the *TimeToRepair* delay, and then a Release step immediately after this delay. In both steps we will specify the *RepairPerson* property as the object to seize/release. This new process is shown below:

Notice the symbol to the left of the name indicating that this is an overridden process. We could also restore back to the original process by clicking on Restore in the Process ribbon.

We are now ready to use our new MRI object definition in a model. Click on Model to return to our main model, and drag out a Source, MRI (from the Project Library), and a Sink, and connect them with Paths. Place the default model entity and replace its symbol with a person. Select the MRI to edit its properties and then run. The following shows this simple model in operation:

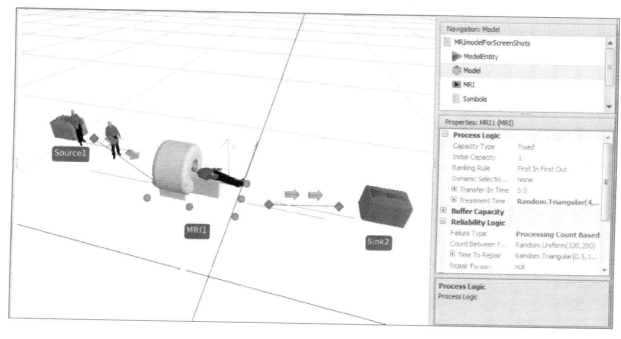

Summary

In this chapter we have reviewed the basic concepts for building object definitions in Simio. This is a key feature of Simio Design and Team Edition because it allows non-programmers to build custom objects that are focused on a specific model or application area.

In our examples here we placed our objects from the Project Library. The other alternative is to build the object definitions in their own project and then load that project as a library from the Project Home ribbon.

With Simio Design Edition and above you often have the choice to make between working with the Standard Library embellished with add-on processes, and creating a new library with custom logic. If you encounter the same applications over and over again it is more convenient to build custom objects and thereby avoid the repeated need for add-on processes. Of course you can also provide support for add-on processes in your own objects by simply defining a property for the process name and passing that property to an Execute step within your process.

With a bit of experience you will find object building to be a simple and powerful capability for your modeling activities.

Glossary

Allocation – Granting ownership of one or more capacity units to an object.

Assign – Defining values for State Variables

Associated Object – In the context of Processes and Tokens, the associated object is the object (if any) that triggers the process. This is typically, but not always, an entity. Also see Parent object.

Batch – A group of entities associated with another entity (e.g. boxes loaded on a pallet)

Buffer Station – A capacity constrained location where objects wait until conditions are met to proceed to their next location.

Capacity – The number of units of a resource that are available to be seized. Once all of the units are seized, no other object can seize this object until the units have been returned. During this time the requester is placed in the Allocation Queue.

Confidence Level – A percentage used to express the how likely the Confidence Interval is to contain the true population mean. For given data, a larger Confidence Level will result in a larger the spread/Half Width

Confidence Interval – A range estimate of a response value (e.g., an estimate of WIP is (352.4 ± 22.7)).

Control – An item whose value can be changed across Scenarios in the Experiment. Controls are added by adding a Reference Property to the model.

Delay – The amount of time that an object suspends all future actions, usually signifying some sort of action is taking place during this time (e.g. a processing time).

Destination – A destination is a node name and can be set using a SetNode step.

Destroy – Remove an object run space from the system during a run.

Discrete Event Simulation (DES) – A type of simulation modeling that deals with events that happen in a chronological sequence.

Distribution – A common alternative to using raw process data is to instead use a tool to fit the data to a probability distribution from which realistic samples of the data can be stochastically generated. Simio contains 20 distributions that may be accessed using the "Random." Keyword.

Element – Things that are capable of changing state over time. For example, the number of objects inside a Server doesn't change; the number of objects in the Processing Station changes. The server just dictates how the objects get into the Station. Types of Elements include:

- Activity – Performed by an operation that is owned by an object
- Batch Logic – Used with Batch step to match multiple entities together and attach the batched group to a parent entity
- Failure – Used to define a failure mode
- Material – Used with consume and produce steps to represent using and ordering defined amounts of a material in a process
- Monitor – Watches a state variable and fires an event when there is a discrete change or if it crosses a certain threshold level
- Network – A collection of Links that certain Entities or Transporters can travel on, while others cannot. A link may be on multiple networks

- Operation – Sequence of Activities that are performed over time
- Output Statistic – An expression that can be optionally recorded at the end of each replication of a simulation
- Routing Group – Used with a Route step to route an object to a destination based on a list of candidate nodes
- State Statistic – Used to record time-persistent statistics on a state variable
- Station – Capacity constrained location where an object can reside. Examples include Processing, Input Buffer, Output Buffer, and Ride Station
- Storage – A logical queue used to create a list of one or more objects
- Tally Statistic – Tallies observational statistics that are recorded using a Tally Step
- Timer – Fires a stream of events according to a specified interval type

Entity – Part of an object model and can have its own intelligent behavior. They can make decisions, reject requests, decide to take a rest, etc. Entities have object definitions just like the other objects in the model. Entity objects can be dynamically created and destroyed, moved across a network of Links and Nodes, move through 3D space, and move into and out of Fixed objects. Examples of Entity objects include customers, parts or work pieces.

Event – A notification that can be given by one object and responded to by several. It alerts other objects that an action has occurred.

Experiment – Part of the project that is used for output analysis. The user defines one or more sets of inputs/outputs (Scenarios) and runs multiple replications to get statistically valid results from which to draw conclusions.

Free Space – The "space" in the Facility view that is represented only by xyz coordinates. Objects can move through free space as an alternative of moving across links in a network.

Half Width – Half of the confidence interval. CI = Average ± Half Width. The smaller the Half Width the smaller the variability.

Learning Mode – A display mode that temporarily hides intermediate and advanced features to facilitate learning by allowing users to concentrate on the basic modeling constructs.

Library – A collection of objects/models that can be used to create a model.

Input Analysis – Distribution-fitting software will automatically help you determine which probability distribution best represents your data. ExpertFit and Stat::Fit are two programs recommended because they have specific support for Simio distributions.

Link – Pathways for entity/transporter movement. There are four types of links:

- Connector – Used to define a non-passing pathway between two node locations where travel time is instantaneous. Entities will never actually enter, move across or exit a connector, but will always instead immediately do a node-to-node direct transfer.
- Conveyor – Represents either accumulating or non-accumulating conveyors and supports both fixed and variable spacing.
- Path – Used to define a pathway between two node locations where the travel time is determined by the path length and a traveler's speed.
- Time Path – Used to define a pathway between two node locations where the travel time is user specified.

Member Entity – An entity that is temporarily associated with a Parent entity, such as the members might be boxes on the pallets represented by a parent.

Model – A representation of real world object or collection of objects that interact with each other. Models are usually used to make decisions and are defined by their properties, states, events, external view, and logic. A model is an object that is executable.

Network – A collection of Links that certain Entities or Transporters can travel on, while others cannot. A link may be on multiple networks

Object – Defines data, logic, behavior, view, events, and interaction with other objects.

Object Hierarchy:

- Object Definition – A collection of Properties, States, Events, Elements, and other characteristics that defines the logic and behavior of an object and how it interacts with other objects. Every model is an object definition that can be potentially used in another model.
- Object Instance – An object definition that is placed into Simio's Facility View. It carries unique property values to identify it and customize its behavior (e.g. an entity's target speed).
- Object Run Space – Objects that are created during a run that carry with them states to record system changes (e.g. an entity's current speed). These object run spaces refer back to the object instance for their properties and to the object definition for their logic and behavior.

Optimization – The process of defining and evaluating experiment scenarios to determine the overall "best" scenario. OptQuest is a tool that is highly integrated with Simio that will automatically optimize against a Single Objective, Multi-Objective Weighted, or Pattern Frontier.

Parent Entity – An entity that is temporarily associated with a Member entity, such as the members might be boxes on the pallets represented by a parent.

Parent Object – In the context of Processes and Tokens, the parent object is the object in which the process is defined. Also see Associated Object.

Passing – Allowing one object to travel through another object on a link.

Pivot Grid (Pivot Table) – A type of output report that can be easily filtered, sorted, and categorized to quickly generate a custom report showing exactly the data desired. Reports can be named and reused.

Priority – Numeric property for all Entities used to make Ranking Rules or Dynamic Selection Rules to order how entities are to be moved throughout a model.

Process – A sequence of commands that dictate the behavior of an object. Specified graphically by placing Steps in a flow process diagram.

Project – A collection of Models and Experiments that are used to represent a system. A project can be loaded as a library.

Property – Input parameters that are associated with the object that define exactly how the object interacts with other objects. Ex – Processing Time, Selection Rule, Priority, Transfer-In Time

Pseudo Random – Something that appears to be random, but is not. DES random number generators are pseudo random in that they have predictability. It repeats the same random number stream every time you press "Run" and can be used again and again, useful for testing and fixing models.

Queue – A collection of objects waiting for an event to occur that will allow them to enter a station, seize an object, enter a link, etc.

Release – Freeing one or more units of the reserved object to allow seizing by other objects.

Random Number Generator – Simio uses the Mersenne Twister Random Number Generator. This is a simple, extremely fast random number generator with a period of 2^19937-1 (i.e. it's huge!).

Random Number Stream – Simio supports an unlimited number of random number streams. All distributions have an optional extra parameter at the end to specify a stream number.

Ranking Rule – A mechanism to control selection of an item from a queue, for example allocating a resource that has just been released. A Static ranking rule physically sorts entities as they are added to the queue. A Dynamic ranking rule examines every entity each time a decision is made to dynamically select the best of all waiting entities.

Reference Property – A mechanism to specify a property value that is set via model properties, object properties, and experiment controls. You may right-click on a property and associate it with a new or existing reference property. They are configured on the Definitions > Properties panel.

Replication – Running a scenario one time, usually run multiple replications per scenario. The more replications the more accurate the statistics.

Resource – An object that represents a system constraint and can be seized and released. Any object can be a resource. Resources may be given intelligent behavior and interact with other objects.

Response – An expression whose value is used to compare the performance between scenarios. These values often also exist in the detailed reports but are identified as a response to indicate their use as a Key Performance Indicator (KPI) and to allow more detailed analysis in the Response Results.

Scenario – A set of input values to be run in an experiment, usually a possible future state of the system being evaluated.

Selection Weight – Property of all links that determines the likelihood of choosing that particular link as an outbound link from an object. Maybe be a simple number or a complex expression.

Seize – To reserve one or more units of an object for use by the requesting object. While an object is seized no other object can take possession of that object.

Server – An object that consists of three stations: Input Buffer, Processing, and Output Buffer. Usually represents a location that has a capacitated process.

SimBit – A brief, well documented, example of solving a specific modeling issue. Basic and Advanced SimBit search engines are provided via the Support Ribbon or descriptive information is in Help.

State – Dynamic values that may change as the model executes. There are four types of states:
- Discrete State – represents a numeric value that may change instantaneously at discrete event times
- Level State – represents a variable that has a numeric value that may change continuously over time based on the value of the rate.
- List State – represents an integer value corresponding to one of several entries from a string list.
- Movement State – represents a variable that has a numeric value that may change continuously over time based on the current value of the rate and acceleration.

Step – Used to define logic within a process. Each step performs an action (such as seize, delay, decide, or wait).

Stochastic Process – A process whose behavior is non-deterministic, meaning that there is randomness in the process.

Symbol – A drawing or representation, often in 3D, that is used to represent a fixed or dynamic model object, or model background to aid in the attractiveness and communication of the model. Symbols can be found in the built-in library, the Trimble (Google) 3D Warehouse, or imported from DXF and other file formats.

Table – A table is a set of rows and columns to hold data during a run. It may be relational and may contain special columns like times or destinations in addition to general model data. Tables may be imported and exported. An object may be associated with a specific table and row using the SetRow Step.

Token – Executes the steps in a process flow. A Token may have one or more user defined states that carry information from step to step.

Transporter – A transporter object is a special type of entity that can pickup entity objects at a location, carry those entities through a network of links or free space, and then drop the entities off at a destination.

Validation – Determining if the model is an accurate representation of the real system – if the model's output matches the real system sufficiently to meet the model objectives.

Variability – A measure of how spread the observed data values are. A key cause of variability is the randomness incorporated into typical simulation models.

Vehicle – A transporter object that represents a machine. A vehicle has the possibility of failures representing breakdowns.

Verification – Determining if the model is built correctly - i.e. with the correct input logic, parameters, and values.

Warm up Time – A time specified on Experiment Properties during which statistics are ignored to eliminate startup bias.

Worker – A transporter object that represents a human. A worker has the ability to follow a work schedule.

More Information

Technical Support

The best way to get support is to use the **Simio User's Forum** found at www.simio.com/forums, or contact us at support@simio.com using one of the links on the Support Ribbon. Sign up on the user's forum as a **Simio Insider** to get full access to find product information, a place to post problems and questions, and the opportunity to engage in discussions with other users and the Simio team.

You can also find Simio User's Groups on:
- **LinkedIn** (www.linkedin.com/groupInvitation?groupID=55167) and
- **Facebook** (www.facebook.com/groups/13863832711).

Please visit our simulation community resources web page (www.simio.com/resources.html) for additional information.

The **Support Ribbon** (the Support tab) of the Simio software contains shortcuts to the above, plus much more. Don't miss the links to our on-line documentation, training, videos, downloads, and other helpful resources. Among other things, you will find the **Introduction to Simio** video series that is the companion to this book. You will also find two other comprehensive video series to help you advance your Simio simulation and scheduling skills. And all of these resources are free, just a few clicks away!

Unfortunately we cannot provide direct support to students. We encourage students to work through their instructors, or to engage the broader user community using the Simio Insiders User Forum.

More Information

Please consult the **Support Ribbon** of the Simio software for a description of more available resources and how to contact us with ideas, questions, or problems.

While Simio includes comprehensive help available at the touch of "F1" in the product, if you prefer a printable version, you will find a link to the **Simio Reference Guide.pdf** under Books on the Support ribbon, as well as links to our increasing number of **Training Videos**.

Using This Material in Support of Teaching

While this document is protected by copyright, we support and encourage its use in Simio education. For educational purposes, you are free to reuse this material and excerpts from it in derivative works as long as the original source is cited.

Appendix 1: Simio and Simulation: Modeling, Analysis, Applications

The following chapter is included by permission from:

Kelton, W. D., J. S. Smith, and D. T. Sturrock. 2011. *Simio and Simulation: Modeling, Analysis, Applications.* 2nd ed. New York: McGraw-Hill, Inc

Copyright © 2011 by W. David Kelton, Jeffrey S. Smith, and David T. Sturrock. All rights reserved.
ISBN-13: 978-0-9-8297825-2
ISBN-10: 0-98-297825-1

The attached is an early preview of the 3rd edition book to be available in late 2013. We encourage you to obtain the entire book, available in both printed and e-book form, for a much more thorough coverage of Simio and simulation.

Chapter 1

Introduction to Simulation

Simulation has been in use for over 40 years, but rather than being "over the hill," it's just moving into its prime. Gartner (www.gartner.com) is a leading provider of technical research and advice for business. In a recent analysis, Gartner [11] identified the top ten strategic technologies for 2010 and ranked *Advanced Analytics*, including simulation, as number two:

> "Optimization and simulation is using analytical tools and models to maximize business process and decision effectiveness by examining alternative outcomes and scenarios, before, during and after process implementation and execution. This can be viewed as a third step in supporting operational business decisions. Fixed rules and prepared policies gave way to more informed decisions powered by the right information delivered at the right time, whether through customer relationship management (CRM) or enterprise resource planning (ERP) or other applications. The new step is to provide simulation, prediction, optimization and other analytics, not simply information, to empower even more decision flexibility at the time and place of every business process action. The new step looks into the future, predicting what can or will happen."

Simulation-related advancements in hardware and software over the last decade have been dramatic. Powerful personal computers now provide processing power unheard of even a few years ago. Advances in user interfaces and product design have resulted in software that's significantly easier to use, lowering the expertise required to use simulation effectively. Breakthroughs in object-oriented technology provide significantly improved modeling flexibility and allow accurate modeling of highly complex systems. Hardware, software, and publicly available symbols allow even novices to produce simulations with compelling 3D animation to support communication between people of all backgrounds. These innovations and other developments are working together to propel simulation into a new position as a critical technology.

In this book we hope to open up the world of simulation to you, providing exposure to general simulation technology and success skills, as well as a practical introduction to a state-of-the-art simulation package.

1.1 About the Book

This book is divided into three parts. The first part, **Simulation Concepts**, encompasses Chapters 1-4. This part is intended to provide a sound basis in the underlying concepts before introducing any software-specific concepts. Chapter 1, *Introduction to Simulation*, covers typical simulation applications, how to identify an appropriate simulation application, and how to carry out a simulation project. Chapter 2, *Basics of Queueing Theory*, introduces the concepts of queueing theory, its strengths and limitations, and in particular how it can be used to help validate components of later simulation modeling. Chapter 3, *Approaches to Simulation*, introduces some of the technical aspects and terminology of simulation, delineates the different kinds of simulation, then illustrates this by working through a manual simulation and two spreadsheet simulations. Chapter 4, *Input Analysis*, discusses different types of inputs to simulations, methods for converting observed real-world data into something useful to a simulation project, and generating the appropriate random quantities needed in most simulations.

The second part of the book, **Simulation Modeling With Simio**, is composed of six chapters. This part introduces more detailed simulation concepts illustrated with numerous examples implemented in Simio. Rather than breaking up the technical components (like validation, and output analysis) into separate chapters, we look at each example as a mini project and introduce successively more concepts with each project. This approach provides the opportunity to learn the best overall practices and skills at an early stage, and then reinforce those skills with each successive project.

Chapter 5, *First Simio Models*, starts with a brief overview of Simio itself, and then directly launches into building a single-server queueing model in Simio. The primary goal of this chapter is to introduce the simulation model-building process using Simio. While the basic model-building and analysis processes themselves aren't specific to Simio, we'll focus on Simio as an implementation vehicle. This process not only introduces modeling skills, but also covers the statistical analysis of simulation output results, experimentation, and model verification. That same model is then reproduced using lower-level tools to illustrate another possible modeling approach, as well as to provide greater insight into what's happening "behind the curtain." The chapter continues with a third, more interesting model of an ATM machine and introduces additional output analysis using Simio's innovative SMORE plots. The chapter closes with some additional discussion of output analysis outside of Simio, as well as adding basic 3D animation to the model.

The goal of Chapter 6, *Intermediate Modeling With Simio*, is to build on the basic Simio modeling-and-analysis concepts presented earlier so that we can start developing and experimenting with models of more realistic systems. We'll start by discussing a bit more about how Simio works and its general framework. Then we'll build an electronics-assembly model and successively add additional features, including modeling multiple processes, conditional branching and merging, etc. As we develop these models, we'll continue to introduce and use new Simio features. We'll also resume our investigation of how to set up and analyze sound statistical simulation experiments, this time by considering the common goal of comparing multiple alternative scenarios. By the end of this chapter, you should have a good understanding of how to model and analyze systems of intermediate complexity with Simio.

Chapter 7, *Working With Model Data*, takes a wider view and examines the many types of data that are often required to represent a real system. We'll start by building a simple emergency-department (ED) model, and will show how to meet its input-data requirements using Simio's data-table construct. We'll successively add more detail to the model to illustrate the concepts of sequence tables, relational data tables, arrival tables, and importing and exporting data tables. We'll continue enhancing the ED model to illustrate work schedules, rate tables, and function tables. The chapter ends with a brief introduction to lists, arrays, and changeover matrices. After completing this chapter you should have a good command of the types of data frequently encountered in models, and the Simio choices for representing those data.

Animation and Entity Movement, Chapter 8, discusses the enhanced validation, communication, and credibility that 2D and 3D animation can bring to a simulation project. Then we explore the various animation tools available, including background animation, custom symbols, and status objects. We'll revisit our previous electronics-assembly model to practice some new animation skills, as well as to explore the different types of links available, and add conveyors to handle the work flow. Finally, we'll introduce the Simio Vehicle and Worker objects for assisted entity movement, and revisit our earlier ED model to consider staffing and improve the animation.

Chapter 9 is *Advanced Modeling With Simio*. We start with a simpler version of our ED model, with the goal of demonstrating the use of models for decision-making, and in particular simulation-based optimization. Then we'll introduce a new pizza-shop example to illustrate a few new modeling constructs, as well as bring together concepts that were previously introduced. A third and final model, an assembly line, allows study of buffer-space allocation to maximize throughput.

Chapter 10, *Customizing and Extending Simio* starts with some slightly more advanced material — it builds on the prior experience using add-on processes to provide guidance in building your own custom objects and libraries. It includes examples of building objects hierarchically from base objects, and sub-classing standard library objects. This chapter ends with an introduction to Simio's extendability through programming your own rules, components, and add-ons to Simio.

The third and final part of the book is **Case Studies Using Simio**. Chapter **??**, *Introductory Cases*, provides four somewhat small but realistic case studies to allow you to practice your modeling skills. Chapter **??**, *Advanced Cases*, provides two larger and more challenging case studies for more advanced study.

1.2 Systems and Models

A *system* is a very broad term used to describe a set of related components that together work toward some purpose. A system might be something as simple as a waiting line at an automated teller machine (ATM), or as complex as a complete airport or a worldwide distribution network. In any such system, be it existing or merely contemplated, it's natural and sometimes even essential to understand how it will behave and perform under various configurations and circumstances.

If the system already exists, sometimes you can gain the necessary understanding by careful observation. One drawback of this approach is that you may need to watch the real system a long time in order to observe the particular conditions of interest even once, let alone making enough observations to reach reliable conclusions. And of course, for some systems (say that worldwide distribution network), it's hard to find the vantage point from which you can observe the entire system at an adequate level of detail. Additional problems arise when you want to study changes to the system. In some cases it may be easy simply to make the change in the real system — for example, add a temporary second person to a shift to observe the impact. But in many cases this is just not practical — consider the investment required to evaluate whether you should use a standard machine that costs $300,000 or a high-performance machine that costs $400,000. And finally, if the real system doesn't yet exist, no observation is possible at all.

For all the reasons above, we often choose to use some sort of *model* of the system to gain understanding. There are many types of models, each with its own advantages and limitations. *Physical models*, such as a model of a car or airplane, can provide both a sense of reality as well as interaction with the physical environment, as in wind-tunnel testing. There are many different types of *analytical models* that use mathematical representations to facilitate understanding — these can be quite good in specific problem domains, but available domains are often limited. Simulation is yet another modeling approach that has much broader applicability.

Computer simulation is the imitation of the operation of a system and its internal processes, usually over time, and in appropriate detail to draw conclusions about the system's behavior. Simulation models are most often created using software designed to represent common system components, their behavior and time-based interactions, and to record an artificial "history" of a model run as well as summaries and inferences about system characteristics. Simulation is often used for both predicting the effect of changes to existing systems, as well as predicting the performance of new systems. Simulations are frequently used in the design, emulation, and operation of systems.

Simulations may be stochastic or deterministic. In a *stochastic* simulation (the most common), randomness is introduced to represent the variation found in most systems. For example, the results of activities involving people (time to complete a task, quality level) always vary, external inputs (customers, materials) vary, and exceptions (failures) occur. *Deterministic* models have no variation. These are rare in design applications, but more common in model-based decision support such as scheduling and emulation applications. Section 3.1.3 discusses this further.

There are two main types of simulation, *discrete* and *continuous*. The terms discrete and continuous refer to the changing nature of the states that describe the system. Some states (e.g., the length of a queue, status of a worker) can change only at discrete points in time (called *event times*). Other states (e.g.,

1.2. SYSTEMS AND MODELS

pressure in a tank, temperature in an oven) can change continuously over time. Some systems are purely discrete or continuous, while others have both types of states present. In Section 3.1.2 we discuss this further, and give an example of a continuous simulation.

Continuous systems are defined by *differential equations* that specify the rate of change — simulation software uses numerical integration to generate a solution for the differential equations over time. *System dynamics* is a graphical approach for creating simple models using the same underlying concept, and is often used to model population dynamics, market growth/decay, etc.

There are four discrete modeling paradigms that have evolved. *Events* model the points in time when the system state can change (e.g., a customer arrival or departure). *Processes* model a sequence of actions that take place over time (a part in a manufacturing system seizes a worker, delays by a service time, then releases the worker). *Objects* describe the model from the point of view of the facility. *Agent-based modeling* (ABM) is a special case of objects — the system behavior emerges from the interaction of a large number of autonomous intelligent objects (soldiers, firms in a market, infected individuals in an epidemic, etc.). The distinction between these paradigms is somewhat blurred because some modern packages incorporate multiple paradigms. Simio is a multi-paradigm modeling tool that combines all these paradigms into a single framework. You can use a single paradigm, or combine multiple paradigms in the same model. Simio combines the ease of objects with the flexibility of processes.

Simulation has been applied to a huge variety of settings. Here are just a few samples of areas where simulation has been used to understand and improve the system effectiveness:

Airports: Parking-lot shuttles, ticketing, security, terminal transportation, food court, baggage handling, gate assignment, airplane de-icing.

Hospitals: Emergency department, disaster planning, ambulance dispatching, regional service strategies, resource allocation.

Ports: Inbound traffic, outbound traffic, port management, container storage, capital investments, crane operation.

Mining: Material transfer, labor transportation, equipment allocation, bulk material mixing.

Amusement parks: Guest transportation, ride design/startup, waiting lines, ride staffing, crowd management.

Call centers: Staffing, skill-level assessment, service improvement, training plans, scheduling algorithms.

Supply chains: Risk reduction, reorder points, production allocation, inventory positioning, transportation, growth management, contingency planning.

Manufacturing: Capital-investment analysis, line optimization, product-mix changes, productivity improvement, transportation, labor reduction.

Military: Logistics, maintenance, combat, counterinsurgency, search and detection, humanitarian relief.

Telecommunications: Message transfer, routing, reliability, network robustness to outages or attacks.

Criminal-justice system: Probation/parole operations, prison utilization and capacity.

Emergency-response system: Response time, station location, equipment levels, staffing.

Public sector: Allocation of voting machines to precincts.

Customer service: Direct-service improvement, back-office operations, resource allocation, capacity planning.

Some people still think of simulation as a tool only for manufacturing, but that's obviously not the case. The domains and applications of simulation are wide ranging and virtually limitless.

1.3 Randomness and the Simulation Process

1.3.1 Randomness in Simulation and Random Variables

While there are examples of simulation modeling using only deterministic values, the vast majority of simulation models incorporate some form of randomness that is inherent in the systems being modeled. Typical random components include processing times, service times, customer/entity arrival times, transportation times, machine/resource failures and repairs, etc. For example, if you plan head to the drive-thru at a local fast-food restaurant for a late-night snack, you generally do not know exactly how long it will take you to get there, how many other customers may be in front of you when you arrive, how long it will take to be served, etc. While we may be able to *estimate* these values based on prior experience or other knowledge, we generally cannot predict them with certainty. Using deterministic estimates of these stochastic values in models can result in invalid (generally overly optimistic) performance predictions. However, incorporating these random components in standard analytical models can be difficult or impossible. Using simulation, on the other hand, makes inclusion of random components quite easy and, in fact, it is precisely its ability to easily incorporate stochastic behavior that makes simulation such a popular modeling and analysis tool. This will be a fundamental theme throughout this book.

Randomness in simulation models is generally expressed using *random variables*. As such, comprehensive understanding and use of random variables is fundamental to simulation modeling and analysis. We assume that you are familiar with the basic concepts of random variables (see [43], [38] if you need a refresher). At its most basic, a random variable is a function whose value is determined by the outcome of an experiment. That is, we do not know the value until after we perform the experiment – in the simulation context, this often

Table 1.1: Probability mass (PMF) and density (PDF) functions for random variables.

Discrete Random Variables	Continuous Random Variables
$p(x_i) = Pr(X = x_i)$ $$F(x) = \sum_{\substack{\forall i \ni \\ x_i \leq x}} p(x_i)$$	$f(x)$ has the following properties: 1. $f(x) \geq 0 \; \forall$ real values, x 2. $\int_{-\infty}^{\infty} \mathrm{d}x = 1$ 3. $pr(a \leq x \leq b) = \int_a^b f(x)\mathrm{d}x$

means running the simulation model. The probabilistic behavior of a random variable, X, is described by its distribution function (or *cumulative distribution function*, CDF), $F(x) = Pr(X \leq x)$. For discrete random variables, we are also interested in the probability mass function, $p(x_i)$, and for continuous random variables, we are also interested in the probability density function, $f(x)$ (see Table 1.1). Once we've characterized a random variable X, we are generally interested in metrics such as the expected value ($E[X]$), the variance (Var$[X]$), and various other characteristcs of the distribution such as quantiles, symmetry/skewness, etc. In many cases, we must rely on the sample statistics such as the sample mean, $\bar{X}$, and sample variance, $S^2(X)$, as we cannot feasibly characterize the corresponding population parameters. Determining the appropriate sample sizes for these estimates is important – unlike many other experimental methods, in simulation analysis, we can often control the sample sizes to meet our needs.

From the simulation *input* side, we're interested in characterizing random variables and generating samples from the corresponding distributions and from the *output* side we're interested in analyzing the characteristics of the distributions (i.e., mean, variance, percentiles, etc.) based on observations generated by the simulation. For example, consider a model of a small walk-in healthcare clinic. The system inputs include the patient arrival times and the care-giver diagnosis and treatment times - all of which are random variables (see Figure 1.1 for an example). In order to simulate the system, we need to understand and generate observations of these random variables as inputs to the model. Often, but not always, we have data from the "real" system that we use to characterize the input random variables. Typical outputs of interest include the patient waiting time, time in the system, and the care-giver and space utilizations. The simulation model will generate observations of these random variables. By controlling the execution of the simulation model, we can use the generated observations to characterize the outputs of interest. In the following section, we will discuss in the input and output analysis processes in the context of the general simulation process.

1.3.2 The Simulation Process

The basic simulation process is shown in Figure 1.2. Note that the process is

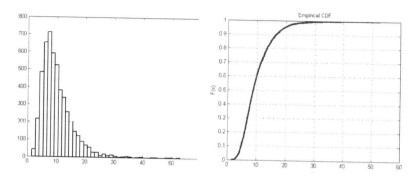

Figure 1.1: Sample patient treatment times and the corresponding empirical CDF.

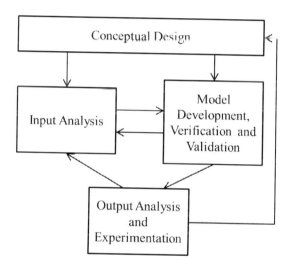

Figure 1.2: The simulation process.

not strictly sequential and will often turn out to be iterative. We will briefly discuss each of these components in the following sections and will develop the topics in detail throughout the book.

1.3.3 Conceptual Design

The conceptual design process involves developing a detailed understanding of the system being modeled and designing a basic modeling approach for creating the simulation model(s). Conceptual design is generally done with pen and paper or on a whiteboard or similar collaboration space that promotes free thinking (i.e., outside of the constraints of the simulation software package that you happen to be using). While it would be nice to have a well-defined process and/or methodology for conceptual design, we do not know of one. Instead, it tends to be an informal process involving "thinking about" and discussing the details of the problem and the potential modeling approaches followed by a systematic detailing of the modeling approach and the application of software-specific details. Note that simulation models are developed for *specific objectives* and an important aspect of conceptual design is ensuring that the model will answer the questions being asked. In general, new simulationists (and new model builders in other domains also) spend far too little time in the conceptual design phase and, instead, tend to jump in and start the model development process (probably because of the cool software like Simio used for model development). Allocating too little time for conceptual design almost always increases the overall time required to for the project.

1.3.4 Input Analysis

Input analysis (which is covered in detail in Chapter yy) involves characterizing the system inputs and developing the computer code for generating observations of the random variables. Luckily, virtually all commercial simulation software (Simio included, of course) has built-in features for generating the observations, so the primary input analysis task involves characterizing the input random variables and specifying corresponding distributions to the simulation software. Often we have sample observations of the data and our task is to "fit" this data to standard or empirical distributions that can be used to generate the samples (as show in Figure 1.1). If we don't have samples, we can use general rules-of-thumb and sensitivity analysis to help with the input analysis task.

1.3.5 Model Development, Verification and Validation

Model development is the "coding" part of the process where the conceptual model is converted into an "executable" simulation model. We don't want to scare anybody off with the term "coding" – most modern simulation packages provide sophisticated graphical user interfaces to support modeling building/maintenance so the "coding" generally involves dragging and dropping model components and filling in dialog boxes and property windows. However, effective model development does require detailed understanding simulation methodology and how the software being used works. The verification and validation steps involve ensuring that the model is "correct." The verification component ensures that the model behaves as the developer intended and the validation component ensures that the model is accurate relative to the actual system being modeled. Note that proving correctness in any but the simplest models will not be possible. Instead, we focus on collecting evidence until we (or our customers) are satisfied. While this may be a bit disturbing to early simulationists, it is reality. Model development, verification, and validation topics are covered starting in Chapter 5 and continuing through the remainder of the book.

1.3.6 Output Analysis and Experimentation

Once a model has been verified and validated, you then exercise the model to glean information about the underlying system. In the above examples, you may be interested in assessing performance metrics like the average time that a patient waits before seeing a care-giver, the 90th percentile of the number of patients in the waiting room, the average number of vehicles waiting in the drive-thru lane, etc. You may also be interested in making design decisions such as the number of care-givers required to ensure that the average patient waits no more than 30 minutes, the number of kitchen personnel to ensure that the average order is ready in 5 minutes, etc. Assessing performance metrics and making design decisions using a simulation model involves output analysis and experimentation. Output analysis is essentially taking the individual observations generated by the simulation, characterizing the underlying random variables (in a statistically valid way), and drawing inferences about the system being modeled. Experimentation involves systematically varying the model inputs and model structure to investigate alternative system configurations. We have chosen to not have a dedicated chapter on output anasyis and, instead, output analysis topics are spread throughout the modeling chapters (x, y, z).

1.4 When to Simulate (and When Not To)

Simulation of complicated systems has become quite popular. One of the main reasons for this is embodied in that word "complicated." If the system of interest were actually simple enough to be validly represented by an exact analytical model, simulation wouldn't be needed, and indeed shouldn't be used: we should instead use such exact analytical methods like queueing theory, probability, or maybe even just simple algebra or calculus. Simulating a simple system for which we can find an exact analytical solution can only add noise, i.e. uncertainty, to the results, making them less precise.

But in reality we quickly get out of the realm of such very "simple" models since, well, the world tends to be a complicated place. And if we're serious about building a *valid* model of a complicated system, that model will itself likely be fairly complicated and not amenable to a simple analytical analysis. We could go ahead and build a simple model of a complicated system with the goal of preserving our ability to get an exact analytical solution, but the resulting model would probably be *overly* simple (simplistic, even), and we'd be left wondering if such a model validly represents the system. We may be able to get a nice, clean, exact, closed-form analytical solution to our simple model, but because we probably made a lot of simplifying assumptions (some of which might be quite questionable in reality) to get to our analytically-tractable model, it's hard to say of what, exactly, in reality do we have a solution — sure, it's a solution to the *model*, but that model might not bear much resemblance to reality so we may not have a solution to the *problem*.

And it's hard, maybe impossible, to quantify or measure just *how* unrealistic a model is; it's not even clear that asking such a question means much. On the other hand, if we don't concern ourselves with building a model that will have an analytical solution in the end, we're freed up to allow things in the model to become as complicated and messy as they need to be in order to mimic the system in a valid way. But, a nice simple analytically-tractable model is no longer available, so we need to turn to simulation, where we simply mimic the complicated system, via its complicated (but realistic) model, numerically on a computer, and watch what happens to the results. Part of this is usually to allow some model inputs to be *stochastic*, i.e., random and represented by "draws" from probability distributions rather than by fixed constant input values, if such is the way things are in reality; this causes the results from our simulation model to be likewise stochastic, and thus uncertain.

Clearly, this uncertainty or imprecision in simulation output is a downside. But, as we'll, see, it's *not* hard to measure the degree of this imprecision, and if we don't like the answer (i.e., the results are too imprecise), we know how to deal with it, and can indeed do so, often by just simulating some more and consuming more computer time. Unlike most statistical sampling experiments, we're in complete control of the "randomness" and numbers of replications, and can use this control to gain any level of precision that we like. Computer time used to be a real barrier to simulation's utility, and in some cases it still can be for tremendously complex models that require hours or maybe days to run once. But for many simulation models, it's now possible with readily available (and relatively cheap) computing power to do enough simulating to get results with imprecision that's both measurable and acceptably low — and with no gnawing doubt about whether our model is so simple as to be unrealistic and not valid.

In years gone by, simulation has sometimes been dismissed as "the method of last resort," or an approach to be taken only "when all else fails" ([51], pp. 887, 890). As noted above, we agree that simulation should not be used if a *valid* analytically-tractable model is available. But in many (perhaps most) cases, the actual system is just too complicated, or does not obey the rules, to allow for an analytically-tractable model of any credible validity to be built and analyzed. In our opinion, it's better to simulate the right model and get an approximate answer whose imprecision can be objectively measured and reduced, than to do an exact analytical analysis of the wrong model and get an answer whose error cannot be even be quantified, a situation that's worse than imprecision.

While we're talking about precise answers, the examples and figures in this text edition were created with Simio Version 4.58. Because each version of Simio may contain changes that could affect low-level behavior (like the processing order of simultaneous events), different versions could produce different numerical output results for an interactive run. You may wonder "Which results are correct?" Each one is as correct (or as incorrect) as the others! In this book you'll learn how to create *statistically* valid results, and how to recognize when you have (or don't have) them. With the possible exception of a rare bug fix between versions, every version should generate *statistically* equivalent (and valid) results for the same model, even though they may differ numerically across single interactive runs.

1.5 Simulation Success Skills

Learning to use a simulation tool and understanding the underlying technology will not guarantee your success. Conducting successful simulation projects requires much more than that. Newcomers to simulation often ask how they can be successful in simulation. The answer is easy: "Work hard and do everything right." But perhaps you want a bit more detail. Let's identify some of the more important issues that should be considered.

1.5.1 Project Objectives

The first question to ask when presented with a simulation is "What are your objectives?" Although it may seem like an obvious question with a simple answer, it often happens that stakeholders don't know the answer. Many projects start with a fixed deliverable date, but often only a rough idea of what will be delivered and a vague idea of how it will be done.

Your first role may be to help clarify the objectives. But before you can help with objectives, you need to get to know the stakeholders. A *stakeholder* is someone who commissions, funds, uses, or is affected by the project. Some stakeholders are obvious — your boss is likely to be stakeholder (if you're a student, your instructor is most certainly a stakeholder). But sometimes you have to work a bit to identify all the key stakeholders. Why should you care? In part because stakeholders usually have differing (and conflicting) objectives.

Let's say that you're asked to model a specific manufacturing facility at a large corporation, and evaluate whether a new $4 million crane will provide the desired results (increases in product throughput, decreases in waiting time, reductions in maintenance, etc.). Here are some possible stakeholders and what their objectives might be in a typical situation:

- Manager of industrial engineering (IE) (your boss): She wants to prove that IE adds value to the corporation, so she wants you to demonstrate dramatic cost savings or productivity improvement. She also wants a nice 3D animation she can use to market your services elsewhere in the corporation.

- Production Manager: He's convinced that buying a new crane is the only way he can meet his production targets, and has instructed his key people to provide you the information to help you prove that.

- VP-Production: He's been around a long time and is not convinced that this "simulation" thing offers any real benefit. He's marginally supporting this effort due to political pressure, but fully expects (and secretly hopes) the project will fail.

- VP-Finance: She's very concerned about spending the money for the crane, but is also concerned about inadequate productivity. She's actually the one who, in the last executive meeting, insisted on commissioning a simulation study to get an objective analysis.

- Line Supervisor: She's worked there 15 years and is responsible for material movement. She knows that there are less-expensive and equally effective ways to increase productivity, and would be happy to share that information if anyone bothered to ask her.

- Materials Laborer: Much of his time is currently spent moving materials, and he's afraid of getting laid off if a new crane is purchased. So he'll do his best to convince you that a new crane is a bad idea.

- Engineering Manager: His staff is already overwhelmed, so he doesn't want to be involved unless absolutely necessary. But if a new crane is going to be purchased, he has some very specific ideas of how it should be configured and used.

This scenario is actually a composite of some real cases. Smaller projects and smaller companies might have fewer stakeholders, but the underlying principles remain the same. Conflicting objectives and motivations are not at all unusual. Each of the stakeholders has valuable project input, but it's important to take their biases and motivations into account when evaluating their input.

So now that we've gotten to know the stakeholders a bit, we need to determine how each one views or contributes to the project objectives and attempt to prioritize them appropriately. In order to identify key objectives, you must ask questions like these:

- What do you want to evaluate, or hope to prove?

- What's the model scope? How much detail is anticipated for each component of the system?

- What components are critical? Which less-important components might be approximated?

- What input data can be made available, how good are they, who will provide them, and when?

- How much experimentation will be required? Will optimum-seeking be required?

- How will any animation be used (animation for validation is quite different from animation presented to a board of directors)?

- In what form do you want results (verbal presentation, detailed numbers, summaries, graphs, text reports)?

One very good way to help identify clear objectives is to design a mock-up of the final report. You can say, *"If I generate a report with the following information in a format like this, will that address your needs?"* Once you can get general agreement on the form and content of the final report, you can often work backwards to determine the appropriate level of detail and address other modeling concerns. This process can also help bring out unrecognized modeling objectives.

Sometimes the necessary project clarity is not there. If so, and you go ahead anyway to plan the entire project including deliverables, resources, and date, you're setting yourself up for failure. Lack of project clarity is a clear call to do the project in phases. Starting with a small prototype will often help clarify the big issues. Based on those prototype experiences, you might find that you can do a detailed plan for subsequent phases. We'll talk more about that next.

1.5.2 Functional Specification

*"If you don't know where you're going,
how will you know when you get there?"*

Carpenter's advice: "Measure twice. Cut once."

If you've followed the advice from Section 1.5.1, you now have at least some basic project objectives. You're ready to start building the model, right? Wrong! In most cases your stakeholders will be looking for some commitments.

- When will you get it done (is yesterday too soon)?

- How much will it cost (or how many resources will it require)?

- How comprehensive will the model be (or what specific system aspects will be included)?

- What will be the quality (or how will it be verified and validated)?

Are you ready to give reliable answers to those questions? Probably not.

Of course the worst possible, but quite common, situation is that the *stakeholder* will supply answers to all of those questions and leave it to you to deliver. Picture a statement like "I'll pay you $5000 to provide a thorough, validated analysis of ... to be delivered five days from now." If accepted, such a statement often results in a lot of overtime to produce a partially complete, unvalidated model that's a week or two late. And as for the promised money ... well, the customer didn't get what he asked for, now, did he?

It's OK for the customer to specify answers to *two* of those questions, and in rare cases maybe even *three*. But you must reserve the right to adjust at least one or two of those answers. You might cut the scope to meet a deadline. Or you might extend the deadline to achieve the scope. Or, you might double both the resources and the cost to achieve the scope and meet the date (adjusting the quality is seldom a good idea).

If you're fortunate, the stakeholder will allow you to answer all four questions (of course, reserving the right to reject your proposal). But how do you come up with good answers? By creating a *functional specification*, which is a document describing exactly what will be delivered, when, how, and by whom. While the details required in a functional specification vary by application and project size, typical components may include:

1. Introduction

    a) Simulation objectives: Discussion of high-level objectives. What's the desired outcome of this project?

    b) Identification of stakeholders: Who are the primary people concerned with the results from this model? Which other people are also concerned? How will the model be used and by whom? How will they learn it?

2. System description and modeling approach: Overview of system components and approaches for modeling them. Including, but not limited to, the following components:

1.5. SIMULATION SUCCESS SKILLS

    a) Equipment: Each piece of equipment should be described in detail, including its behavior, setups, schedules, reliability, and other aspects that might affect the model. Include data tables and diagrams as needed. Where data do not yet exist, they should be identified as such.

    b) Product types: What products are involved? How do they differ? How do they relate to each other? What level of detail is required for each product or product group?

    c) Operations: Each operation should be described in detail including its behavior, setups, schedules, reliability, and other aspects that might affect the model. Include data tables and diagrams as needed. Where data do not yet exist, they should be identified as such.

    d) Transportation: Internal and external transportation should be described in adequate detail.

3. Input data: What data should be considered for model input? Who will provide this information? When? In what format?

4. Output data: What data should be produced by the model? In this section, a mock-up of the final report will help clarify expectations for all parties.

5. Project deliverables: Discuss all agreed-upon project deliverables. When this list is fulfilled, the project is deemed complete.

    a) Documentation: What model documentation, instructions, or user manual will be provided? At what level of detail?

    b) Software and training: If it's intended that the user will interact directly with the model, discuss the software that's required, what software, if any, will be included in the project price quote, and what, if any, custom interface will be provided. Also discuss what project or product training is recommended or will be supplied.

    c) Animation: What are the animation deliverables and for what purposes will the animations be used (model validation, stakeholder buy-in, marketing)? 2D or 3D? Are existing layouts and symbols available, and in what form? What will be provided, by whom, and when?

6. Project phases: Describe each project phase (if more than one) and the estimated effort, delivery date, and charge for each phase.

7. Signoffs: Signature section for primary stakeholders.

At the beginning of a project there's a natural inclination just to start modeling. There's time pressure. Ideas are flowing. There's excitement. It's very hard to stop and do a functional specification. But trust us on this — *doing a functional specification is worth the effort*. Look back at those quotations at the beginning of this section. Pausing to determine where you're going and how you're going to get there can save misdirected effort and wasted time. We recommend that approximately the first 10% of the total estimated project time be spent on creating a prototype and a functional specification. Yes, that means if you expect the project may take 20 days, you should spend about two days on this. As a result, you may well find that the project will require 40 days to finish — certainly bad news, but much better to find out up front while you still have time to consider alternatives (reprioritize the objectives, reduce the scope, add resources, etc.).

1.5.3 Project Iterations

Simulation projects are best done as an iterative process. Even from the first steps. You might think you could just define your objectives, create a functional specification, and then create a prototype. But while you're writing the functional specification, you'll likely discover new objectives. And while you're doing the prototype, you'll discover important new things to add to the functional specification.

As you get further into the project, an iterative approach becomes even more important. A simulation novice will often get an idea and start modeling it, then keep adding to the model until it's complete — and *only then* run the model. But even the best modeler, using the best tools, will make mistakes. But when all you know is that your mistake is "somewhere in the model," it's very hard to find it and fix it. Based on our collective experience in teaching simulation, this is a huge problem for students new to the topic.

More experienced modelers will typically build a small piece of the model, then run it, test it, debug it, and verify that it does what the modeler expected it would do. Then repeat that process with another small piece of the model. As soon as enough of the model exists to compare to the real world, then validate, as much as possible, that that entire section of the model matches the intended system behavior. Keep repeating this iterative process until the model is complete. At each step in the process, finding and fixing problems is much easier because it's very likely a problem in the small piece that was most recently added. And at each step you can save under a different name (like `MyModelV1`, `MyModelV2`, or with full dates and even times appended to the file names), to allow reverting to an earlier version if necessary.

Another benefit of this iterative approach, especially for novices, is that potentially-major problems can be eliminated early. Let's say that you built an entire model based on a faulty assumption of how entity grouping worked, and only at the very end did you discover your misunderstanding. At that point it might require extensive rework to change the basis of your model. However, if you were building your model iteratively, you probably would have discovered your misunderstanding the very first time you used the grouping construct, at which time it would be relatively easy to take a better strategy.

A final, and extremely important benefit of the iterative approach is the ability to prioritize. *For each iteration, work on the most important small section of the model that's remaining.* The one predictable thing about software development of all types is that it almost always takes much longer than expected. Building simulation models often shares that same problem. If you run out of project time when following a non-iterative approach and your model is not yet even working, let alone verified or validated, you essentially have nothing useful to show for your efforts. But if you run out of time when following an iterative approach, you have a portion of the model that's completed, verified, validated, and ready for use. And if you've been working on the highest-priority task at each iteration, you may find that the portion completed is actually enough to fulfill most of the project goals (look up the 80-20 rule or the Pareto principle to see why).

Although it may vary somewhat by project and application, the general steps in a simulation study are:

1. Define high-level objectives and identify stakeholders.

2. Define the functional specification, including detailed goals, model boundaries, level of detail, modeling approach, and output measures. Design the final report.

3. Build a prototype. Update steps 1 and 2 as necessary.

4. Model or enhance a high-priority piece of the system. Document and verify it. Iterate.

5. Collect and incorporate model input data.

6. Verify and validate the model. Involve stakeholders. Return to step 4 as necessary.

7. Design experiments. Make production runs. Involve stakeholders. Return to step 4 as necessary.

8. Document the results and the model.

9. Present the results and collect your kudos.

As you're iterating, don't waste the opportunity to *communicate regularly with the stakeholders*. Stakeholders don't like surprises. If the project is producing results that differ from what was expected, learn together why that's happening. If the project is behind schedule, let stakeholders know early so that serious problems can be avoided. Don't think of stakeholders as just clients, and certainly not as adversaries. Think of stakeholders as partners — you can help each other to obtain the best possible results from this project. And those results often come from the detailed system exploration that's necessary to uncover the actual processes being modeled. In fact, in many projects a large portion of the value occurs before any simulation "results" are even generated — due to the knowledge gained from the early exploration by modelers, and frequent collaboration with stakeholders.

1.5.4 Project Management and Agility

There are many aspects to a successful project, but one of the most obvious is meeting the completion deadline. A project that produces results after the decision is made has little value. Other, often-related, aspects are the cost, resources, and time consumed. A project that runs over budget may be canceled before it gets close to completion. You must pay appropriate attention to completion dates and project costs. But both of those are outcomes of how you manage the day-to-day project details.

A well-managed project starts by having clear goals and a solid functional specification to guide your decisions. Throughout the project, you'll be making large and small decisions, like:

- How much detail should be modeled in a particular section?
- How much input data do I need to collect?
- To which output data should I pay most attention?
- When is the model considered to be valid?
- How much time should I spend on animation? Analysis?
- What should I do next?

In almost every case, the functional specification should directly or indirectly provide the answers. You've already captured and prioritized the objectives of your key stakeholders. That information should become the basis of most decisions.

One of the things you'll have to prioritize is "evolving specifications" or new stakeholder requests, sometimes called "scope creep." One extreme is to take a hard line and say "if it's not in the functional specification, it's not in the model." While in some rare cases this response may be appropriate and necessary, in most cases it's not. Simulation is an exploratory and learning process. As you explore new areas and learn more about the target system, it's only natural that new issues, approaches, and areas of study will evolve. Refusing to deal with these severely limits the potential value of the simulation (and your value as a solution provider).

Another extreme is to take the approach that the stakeholder is always right, and if she asked you to work on something new, it *must* be the right thing to do. While this response makes the stakeholder happy in the short-term, the most likely longer-term outcome is a late or even unfinished project — and a *very* unhappy stakeholder! If you're always chasing the latest idea, you may never have the time to finish the high-priority work necessary to produce any value at all.

The key is to manage these opportunities — that management starts with open communication with the stakeholders and revisiting the items in the functional specification and their relative priorities. When something is added to the project, something else needs to change. Perhaps addressing the new item is important enough to postpone the project deadline a bit. If not, perhaps this new item is more important than some other task that can be dropped (or moved to the "wish list" that's developed when things go better than expected). Or perhaps this new item itself should be moved to the "wish list."

Our definition of *agility* is the ability to react quickly and appropriately to change. Your ability to be agile will be a significant contributor to your success in simulation.

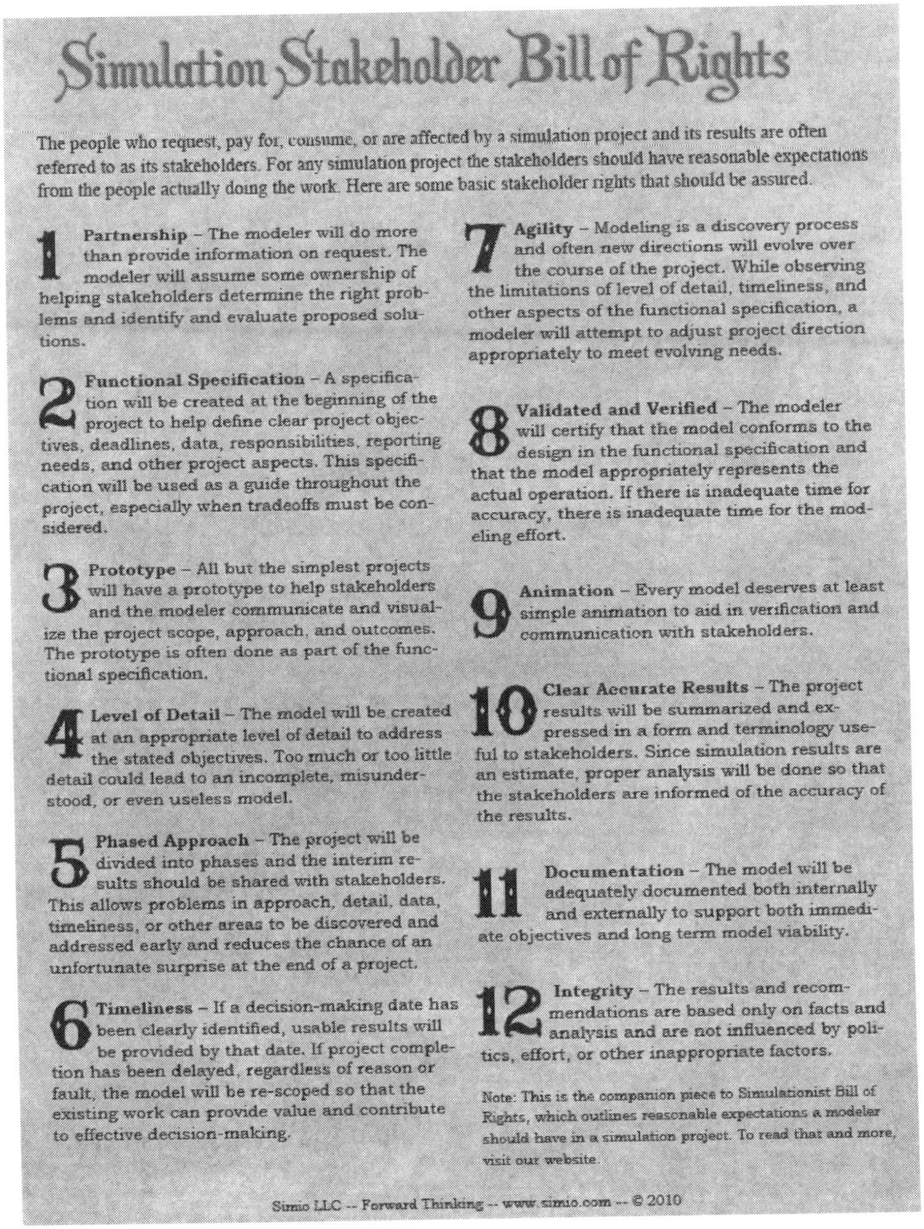

Simulation Stakeholder Bill of Rights

The people who request, pay for, consume, or are affected by a simulation project and its results are often referred to as its stakeholders. For any simulation project the stakeholders should have reasonable expectations from the people actually doing the work. Here are some basic stakeholder rights that should be assured.

1. Partnership – The modeler will do more than provide information on request. The modeler will assume some ownership of helping stakeholders determine the right problems and identify and evaluate proposed solutions.

2. Functional Specification – A specification will be created at the beginning of the project to help define clear project objectives, deadlines, data, responsibilities, reporting needs, and other project aspects. This specification will be used as a guide throughout the project, especially when tradeoffs must be considered.

3. Prototype – All but the simplest projects will have a prototype to help stakeholders and the modeler communicate and visualize the project scope, approach, and outcomes. The prototype is often done as part of the functional specification.

4. Level of Detail – The model will be created at an appropriate level of detail to address the stated objectives. Too much or too little detail could lead to an incomplete, misunderstood, or even useless model.

5. Phased Approach – The project will be divided into phases and the interim results should be shared with stakeholders. This allows problems in approach, detail, data, timeliness, or other areas to be discovered and addressed early and reduces the chance of an unfortunate surprise at the end of a project.

6. Timeliness – If a decision-making date has been clearly identified, usable results will be provided by that date. If project completion has been delayed, regardless of reason or fault, the model will be re-scoped so that the existing work can provide value and contribute to effective decision-making.

7. Agility – Modeling is a discovery process and often new directions will evolve over the course of the project. While observing the limitations of level of detail, timeliness, and other aspects of the functional specification, a modeler will attempt to adjust project direction appropriately to meet evolving needs.

8. Validated and Verified – The modeler will certify that the model conforms to the design in the functional specification and that the model appropriately represents the actual operation. If there is inadequate time for accuracy, there is inadequate time for the modeling effort.

9. Animation – Every model deserves at least simple animation to aid in verification and communication with stakeholders.

10. Clear Accurate Results – The project results will be summarized and expressed in a form and terminology useful to stakeholders. Since simulation results are an estimate, proper analysis will be done so that the stakeholders are informed of the accuracy of the results.

11. Documentation – The model will be adequately documented both internally and externally to support both immediate objectives and long term model viability.

12. Integrity – The results and recommendations are based only on facts and analysis and are not influenced by politics, effort, or other inappropriate factors.

Note: This is the companion piece to Simulationist Bill of Rights, which outlines reasonable expectations a modeler should have in a simulation project. To read that and more, visit our website.

Simio LLC -- Forward Thinking -- www.simio.com -- © 2010

Figure 1.3: Simulation Stakeholder Bill of Rights.

1.5.5 Stakeholder and Simulationist Bills of Rights

We'll end this chapter with an acknowledgement that stakeholders have reasonable expectations of what you will do for them (Figure 1.3). Give these expectations careful consideration to improve the effectiveness and success of your next project. But along with those expectations, stakeholders have some responsibilities to you as well (Figure 1.4). Discussing both sets of these expectations

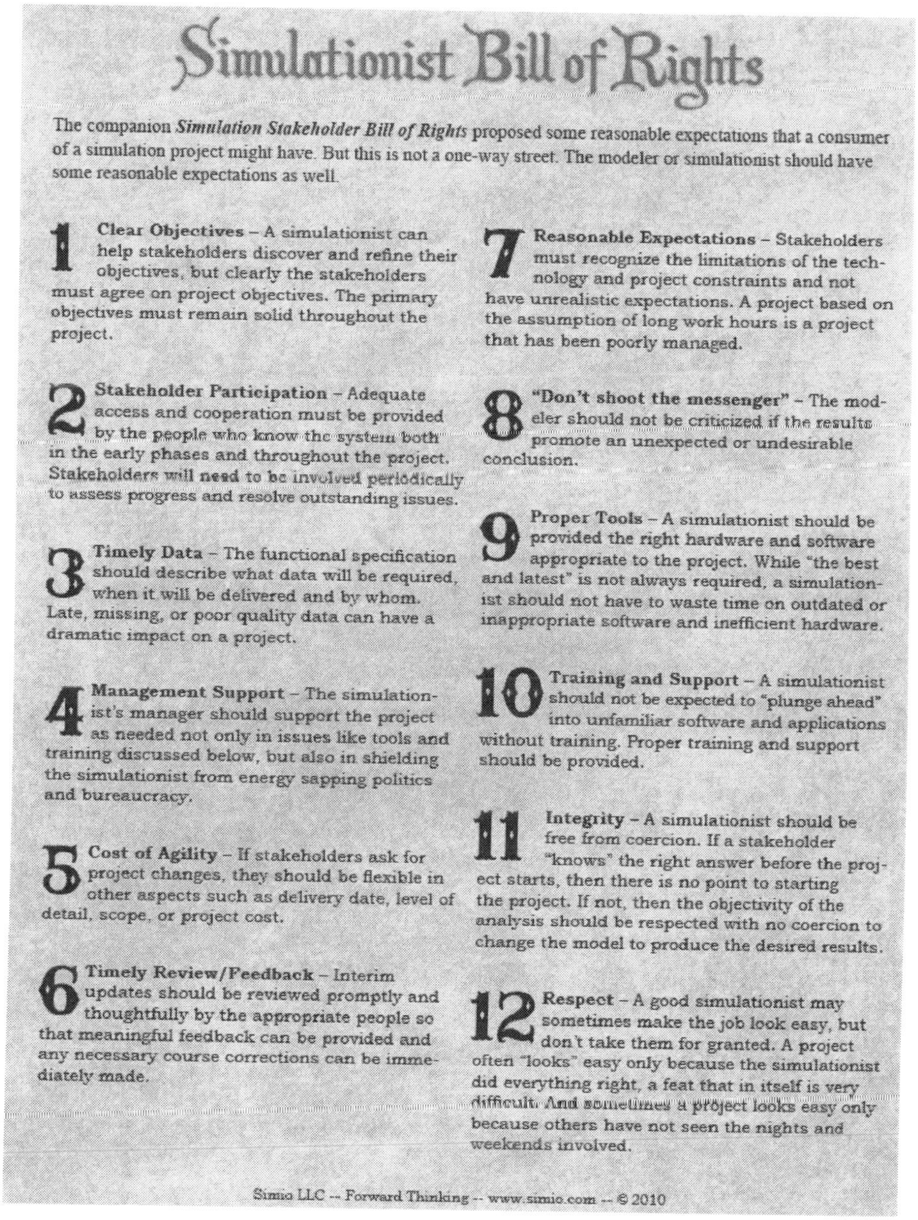

Simulationist Bill of Rights

The companion *Simulation Stakeholder Bill of Rights* proposed some reasonable expectations that a consumer of a simulation project might have. But this is not a one-way street. The modeler or simulationist should have some reasonable expectations as well.

1. Clear Objectives – A simulationist can help stakeholders discover and refine their objectives, but clearly the stakeholders must agree on project objectives. The primary objectives must remain solid throughout the project.

2. Stakeholder Participation – Adequate access and cooperation must be provided by the people who know the system both in the early phases and throughout the project. Stakeholders will need to be involved periodically to assess progress and resolve outstanding issues.

3. Timely Data – The functional specification should describe what data will be required, when it will be delivered and by whom. Late, missing, or poor quality data can have a dramatic impact on a project.

4. Management Support – The simulationist's manager should support the project as needed not only in issues like tools and training discussed below, but also in shielding the simulationist from energy sapping politics and bureaucracy.

5. Cost of Agility – If stakeholders ask for project changes, they should be flexible in other aspects such as delivery date, level of detail, scope, or project cost.

6. Timely Review/Feedback – Interim updates should be reviewed promptly and thoughtfully by the appropriate people so that meaningful feedback can be provided and any necessary course corrections can be immediately made.

7. Reasonable Expectations – Stakeholders must recognize the limitations of the technology and project constraints and not have unrealistic expectations. A project based on the assumption of long work hours is a project that has been poorly managed.

8. "Don't shoot the messenger" – The modeler should not be criticized if the results promote an unexpected or undesirable conclusion.

9. Proper Tools – A simulationist should be provided the right hardware and software appropriate to the project. While "the best and latest" is not always required, a simulationist should not have to waste time on outdated or inappropriate software and inefficient hardware.

10. Training and Support – A simulationist should not be expected to "plunge ahead" into unfamiliar software and applications without training. Proper training and support should be provided.

11. Integrity – A simulationist should be free from coercion. If a stakeholder "knows" the right answer before the project starts, then there is no point to starting the project. If not, then the objectivity of the analysis should be respected with no coercion to change the model to produce the desired results.

12. Respect – A good simulationist may sometimes make the job look easy, but don't take them for granted. A project often "looks" easy only because the simulationist did everything right, a feat that in itself is very difficult. And sometimes a project looks easy only because others have not seen the nights and weekends involved.

Simio LLC -- Forward Thinking -- www.simio.com -- © 2010

Figure 1.4: Simulationist Bill of Rights.

ahead of time can enhance communications and help ensure that your project is successful — a win-win situation that meets everyone's needs. These "rights" were excerpted from the Success in Simulation [49] blog at `www.simio.com/blog` and used with permission. We urge you to peruse the early topics of this non-commercial blog for its many success tips and short interesting topics.

Made in the USA
San Bernardino, CA
28 August 2016